Information Storage

Cornelia S. Große • Rolf Drechsler
Editors

Information Storage

A Multidisciplinary Perspective

Editors
Cornelia S. Große
Group of Computer Architecture
Faculty 3 - Mathematics
and Computer Science
University of Bremen
Bremen, Germany

Rolf Drechsler
Group of Computer Architecture
Faculty 3 - Mathematics
and Computer Science
University of Bremen
Bremen, Germany

Cyber-Physical Systems
DFKI GmbH
Bremen, Germany

ISBN 978-3-030-19264-8 ISBN 978-3-030-19262-4 (eBook)
https://doi.org/10.1007/978-3-030-19262-4

This Springer imprint is published by the registered company Springer Nature Switzerland AG.
The registered company address is: Gewerbestrasse 11, 6330 Cham, Switzerland

Preface

The ability to remember is one of the most important abilities in our life—without it, we would not recognize our parents, our children, our homes. For thousands of years, mankind has been collecting information on how to build houses, on how to cultivate land, on how to raise animals. Each day, new information is acquired, and it is added to our collective knowledge base. Without the ability to store and to retrieve information, our society would not have advanced to its current level. The way we see ourselves, what we are, and what we can be is highly dependent on what we know about ourselves and about the world around us.

Most of what we consider as everyday routine would not be possible had we not acquired and stored the knowledge on how to do it at some point. Each individual stores his or her own experiences, and these memories fundamentally shape our personality.

Moreover, society itself is storing information which is essential in order to maintain and expand our civilization. This second kind of storing information is accomplished in several ways: Within small close-tied communities, like families, information is preserved across generations through interpersonal exchange. On a larger scale, society uses artifacts, pictures in caves, and writings on stones, clay, and parchment, like those seen in museums and archives. Yet there are also television films, books, newspapers, and countless other cultural materials. Parallel to this evolution in our technical possibilities, many of us now have our own "personal museums" stored on computers at home, with hundreds of photos and other personal data files full of things we assume to be important for us. On an interpersonal level, companies and public administration offices utilize large databases interconnected through wires and fiber glass.

In this book, we want to shed light on some of the underlying processes behind these different forms of information management: How does information storage work on these different levels? How can we store information in our brain? What is the impact of new technologies, like computers and robots, on our efficiency of storing information? How is information stored in families and in society? Clearly, "information" is a very broad term with different meanings in different disciplines. Across disciplines it is generally agreed however that information reduces uncertainties and that the ability to store it safely is of vital importance for each of us. In computer science, a typical unit of information is a bit, and any combination of 0 and 1 can represent a multitude of content. In contrast, neuroscientists emphasize the importance of information as sensory inputs that are processed and transformed in the brain. Psychological theories focus more on individual learning and on the acquisition of knowledge. From a sociological perspective, interpersonal processes within groups or society itself—possibly across history—come to the fore.

In this book, experts from a variety of different disciplines contribute to this broad range of topics. To begin with, insights into information storage in the human brain are presented in Chap. 1 "Information Processing and Storage in the Brain." Manfred Fahle provides a review on brain research, pointing out the strengths and weaknesses of human beings. In conjunction with Chap. 2, "Verbal Short-Term Memory: Insights into Human Information Storage" by Michail D. Kozlov, abilities, skills, and characteristics of human beings are considered.

In our everyday life, technology is of major importance, and often we rely on it in order to assist our memory—at least with respect to phone numbers or directions, our smartphones and navigation systems come in very handy. Yet new technologies are emerging every day, transforming the ways in which we handle information. Hence, Chaps. 3 and 4 focus on future developments in computer science. While traditional computers clearly separate between computing components and storing components, in their chapter on "In-Memory Computing: The Integration of Storage and Processing," Saeideh Shirinzadeh and Rolf Drechsler explain how the traditional architecture can be overcome, and how new computers can be designed in a much more efficient way. Another important issue in this area refers to the level of accuracy needed in storing information. In their chapter on "Approximate Memory: Data Storage in the Context of Approximate Computing," Saman Fröhlich, Daniel Große, and Rolf Drechsler reflect on the relevance of accuracy: Complete accuracy is often not of vital importance, but approximations are good enough. However, while certain inaccuracies may be permissible, it is necessary to ensure that they always stay within specified boundaries.

As robots are becoming increasingly prominent, in Chap. 5, "Information System for Storage, Management and Usage for Embodied Intelligent Systems," Daniel Beßler, Asil Kaan Bozcuoğlu, and Michael Beetz focus on methods and insights into enabling a robot to perform everyday activities. These can easily be performed by humans but can be very difficult for robots: If a robot is told to set the table, there are uncertainties, such as what the table might be or the meaning of "setting it," while the robot may also miss unspoken cues contained within such a request.

The last three chapters of this book return to human beings, looking at them as part of a group, as opposed to isolated independent individuals. Since the most prominent, and arguably the most important, group to us is our family, we first look at how information is stored within that social entity. In Chap. 6, "On 'Storing Information' in Families: (Mediated) Family Memory at the Intersection of Individual and Collective Remembering," Rieke Böhling and Christine Lohmeier examine from the point of view of media and communication studies how information is stored and passed on within families. Specifically, they look at families whose grandparent generation abandoned their homeland, so that preserving culture and tradition was somewhat challenging. The issue of culture and information brings us to modern means of information storage and dissemination, like television and other audiovisual sources. Thus, we address the question how television influences information storage within our society. This is done by Berber Hagedoorn in Chap. 7, "Cultural Memory and Screen Culture: How Television and Cross-Media Productions Contribute to Cultural Memory." The issue of storing cultural information through television relates to the issue of preserving media content given unstable formats, changing technologies, and dying out spare parts and expertise. Finally, in Chap. 8, "The Complicated Preservation of the Television Heritage in a Digital Era," Leif Kramp builds a bridge between social science and technical aspects with respect to storage of information.

We would like to express our gratitude to the authors and to the staff members of Springer, especially to Ronan Nugent, for the excellent cooperation and teamwork. Without their commitment, it would not have been possible to compose this book, which is the first one to integrate interdisciplinary perspectives on this highly current and relevant research topic. In this light, with contributions from leading experts in the respective fields, we proudly open up a new view on interdisciplinary studies on information storage.

Bremen, Germany
November 2019

Cornelia S. Große
Rolf Drechsler

Contents

Editors and Contributors

About the Editors

Cornelia S. Große works in the domain of educational psychology and investigates topics of learning and instruction. She studied Psychology at the Albert-Ludwigs-Universität, Freiburg, Germany, and received her diploma in 2001. She continued working there with Alexander Renkl and finished her doctorate on learning with multiple solution methods in 2004. From 2005 to 2009, she worked as a lecturer in the field of educational diagnostics at the University of Bremen, Germany, followed by a postdoc position funded by the University of Bremen (2010–2011), where she continued working on learning with multiple solution methods. From 2012 to 2016, she investigated ways to foster competencies in mathematical modeling as PI in a DFG project. Since 2016, she has worked with Rolf Drechsler at the University of Bremen. Her current research topics are cognitive processes in learning, learning with worked examples, and the acquisition of competencies in mathematics and computer science.

Rolf Drechsler received his diploma and Dr. phil. nat. in Computer Science from the Goethe-Universität Frankfurt am Main, Germany, in 1992 and 1995, respectively. He worked at the Institute of Computer Science, Albert-Ludwigs-Universität, Freiburg, Germany, from 1995 to 2000, and at the Corporate Technology Department, Siemens AG, Munich, Germany, from 2000 to 2001. Since 2001, Rolf Drechsler has been Full Professor and Head of the Group of Computer Architecture, Institute of Computer Science, at the University of Bremen, Germany. In 2011, he additionally became the Director of the Cyber-Physical Systems Group at the German Research Center for Artificial Intelligence (DFKI) in Bremen. His current research interests include the development and design of data structures and algorithms with a focus on circuit and system design. He is an IEEE Fellow.

Contributors

Michael Beetz is Professor of Computer Science at the Faculty for Mathematics & Informatics of the University of Bremen and Head of the Institute for Artificial Intelligence (IAI). He received his diploma in Computer Science with distinction from the University of Kaiserslautern. His MSc, MPhil, and PhD were awarded by Yale University in 1993, 1994, and 1996, respectively, and he obtained his Venia Legendi from the University of Bonn in 2000. He was vice-coordinator of the German cluster of excellence CoTeSys (Cognition for Technical Systems, 2006–2011) and coordinator of the European FP7 integrating project RoboHow (web-enabled and experience-based cognitive robots that learn complex everyday manipulation tasks, 2012–2016), and is the coordinator of the German collaborative research center EASE (Everyday Activity Science and Engineering) since 2017). His research interests include plan-based control of robotic agents, knowledge processing and representation for robots, integrated robot learning, and cognition-enabled perception.

Daniel Beßler is a PhD student under the supervision of Michael Beetz at the Institute of Artificial Intelligence, University of Bremen, Germany. He studied Computer Science at the University of Bremen and obtained his diploma (M. Eng. equivalent) in 2014. As part of his PhD position, he was a member of the European projects RoboHow (web-enabled and experience-based cognitive robots that learn complex everyday manipulation tasks, 2014–2016), and SAPHARI (Safe and Autonomous Physical Human-Aware Robot Interaction, 2014–2015). In addition, he is a member of the ongoing European research project REFILLS (robotics enabling fully-integrated logistics lines for supermarkets, since 2017), and the German collaborative research center EASE (Everyday Activity Science and Engineering) since 2017). One of his latest papers was selected as a finalist for the Best Robotics Paper Award at AAMAS 2018.

Rieke Böhling is a research associate and PhD student at the Centre for Media, Communication and Information Research (ZeMKI) at the University of Bremen, Germany. In her dissertation, she focuses on (mediated) memories of migration in Europe. In particular, she researches how family memories of migration are communicatively constructed at the intersection of personal histories and memories, family memories (where media practices play a role in the process of remembering) as well as mediated discourses and representations of these histories. She is particularly interested in cultural memory, cultural identity, and migration studies. She is assistant managing editor of *VIEW: Journal of European Television History & Culture*. She holds an MA double degree in European Studies: Euroculture from the University of Groningen, the Netherlands and the University of Deusto, Spain.

Asil Kaan Bozcuoğlu graduated in Computer Engineering at Bilkent University, Turkey in 2010. During his MSc studies, he was a member of KOVAN Research Laboratory at Middle East Technical University, Turkey, working on the European FP7 project ROSSI under the supervision of Erol Şahin. He wrote his master's thesis on the cognitive development of grasp behavior in an open-source humanoid robotic platform. He is a PhD student under the supervision of Michael Beetz at the Institute of Artificial Intelligence, University of Bremen. As part of his PhD studies, he was an intern in the JSK Lab at the University of Tokyo under the supervision of Kei Okada, and in April–May 2018 he was a visitor at the Biointelligence Lab at Seoul National University. One of his latest papers was selected as a finalist for the Best Cognitive Robotics Paper Award and the Best Service Robotics Paper Award at ICRA 2018.

Manfred Fahle received his diploma in Biology from the University of Mainz, Germany, in 1975 and his MD from the University of Tübingen, Germany, in 1977. Until 1981 he worked as PhD student and postdoctoral researcher at the Max Planck Institute for Biological Cybernetics. From 1981 to 1988, he was senior physician and a Heisenberg Professor at the University Eye Hospital, Tübingen. Together with T. Poggio, in 1992 he received the Max Planck Prize for Basic Research. From 1994 to 1998 he was Professor of Ophthalmology, Head of "Section Visual Science," University Eye Hospital, Tübingen. In 1996, he was Wiersma Professor at California Institute of Technology. He worked as Head of Department, Dept. Optometry & Visual Science, City University, London, from 1998 to 1999, and subsequently as Visiting Professor, University College, London (Institute of Ophthalmology) until 2001. Since 1999, he has been the Chair of Human Neurobiology, University of Bremen, and Director of the Centre for Cognitive Research.

Saman Fröhlich received his BSc (2011) and MSc (2013) in Computer Science and Engineering at Hamburg University of Technology (TUHH), Germany. He joined the group for Cyber-Physical Systems at the DFKI GmbH, Germany in 2016. Since 2019, he is a member of the Group of Computer Architecture, University of Bremen, Germany, where he is currently pursuing his PhD degree under the supervision of Rolf Drechsler. His research interests include Approximate Computing, Machine Learning, Symbolic Computer Algebra, and Decision Diagrams.

Daniel Große received his Dr.-Ing. in Computer Science from the University of Bremen, Germany, in 2008, where he then worked as a postdoctoral researcher in the Group of Computer Architecture. In 2010, he was a substitute professor for computer architecture at Albert-Ludwigs-Universität, Freiburg, Germany. From 2013 to 2014, he was CEO of the EDA start-up solvertec focusing on automated debugging techniques. Since 2015, he is a senior researcher at the University of Bremen and the German Research Center for Artificial Intelligence (DFKI) Bremen and also the scientific coordinator of the graduate school System Design (SyDe), funded within the German Excellence Initiative. His research interests include verification, high-level languages, virtual prototyping, debugging, and synthesis. In these areas he has published more than 100 papers in peer-reviewed journals and conferences and served in program committees of numerous conferences, such as DAC, ICCAD, DATE, CODES+ISSS, FDL, and MEMOCODE.

Berber Hagedoorn is Assistant Professor in Media Studies at the University of Groningen, Research Centre for Media Studies and Journalism. She specializes in

Screen Culture, storytelling, and working with (digital) audiovisual materials. In her research, she works to improve search, storytelling, and meaning making for different users of audiovisual sources, including media professionals, researchers, and humanities scholars, using methods from digital humanities, data science, textual analysis, production studies, and user studies. She is the Vice-Chair of ECREA's Television Studies section (European Communication Research and Education Association) and organizes cooperation for European research and education into television's history and its future as a multi-platform storytelling practice. She has extensive experience in Media and Culture Studies and Digital Humanities through large-scale European and Dutch best practice projects on digital heritage and cultural memory, including VideoActive, EUscreen, and CLARIAH. More on her work, publications, and network activities can be found at https://berberhagedoorn.wordpress.com.

Michail D. Kozlov obtained his BSc (Hons) in Psychology (2008) and PhD in Psychology (2012), completing his undergraduate and postgraduate education at Cardiff University, UK. Since then he was a research associate at the Leibniz-Institut für Wissensmedien in Tübingen, and the Eberhard Karls Universität Tübingen. He has authored or coauthored several articles in peer-reviewed journals. His research interests include: Computer-Supported Collaborative Learning, Virtual Environments, and Memory. He was the student representative of the British Psychological Society in Cardiff and is currently a full member of the Deutsche Gesellschaft für Psychologie and the Berufsverband Deutscher Psychologinnen und Psychologen. Currently, he is training to become a psychotherapist at the Luisenklinik in Bad Dürrheim.

Leif Kramp is a media, communication, and history scholar. He works as a researcher and research coordinator at the Centre for Media, Communication and Information Research (ZeMKI), University of Bremen, Germany. He has authored and edited various books about the transformation of media and journalism. He is one of the founding members of the German Association for Media and Journalism Criticism (VfMJ) and serves on the directorate of its scholarship program VOCER Innovation Medialab, which promotes early career journalists and media entrepreneurs developing innovative projects. He is also a founding member of the academic initiative "Audiovisual Heritage." He has been a jury member for the German Initiative News Enlightenment (INA) since 2011 and for the #Netzwende-Award for sustainable innovation in the news media since its foundation in 2017.

Christine Lohmeier is a professor in the Department of Communication Studies at the University of Salzburg, Austria. Her research interests are transcultural communication, media in everyday life, memory studies, and qualitative approaches in general and ethnographic research methods in particular. Christine Lohmeier's publications include Cuban Americans and the Miami media (2014) and Memory in a Mediated World (2016, coedited with Andrea Hajek and Christian Pentzold). Christine Lohmeier is cofounder (with Thomas Wiedemann) of the Netzwerk Qualitative Methoden.

Saeideh Shirinzadeh received her BSc and MSc in Electrical Engineering from the University of Guilan, Iran, in 2010 and 2012, respectively. Since 2014, she has been a PhD candidate in the Group of Computer Architecture at the University of Bremen, Germany, and received her PhD in 2018. She was a visiting scholar at the Swiss Federal Institute of Technology (EPFL), Lausanne, Switzerland within the Laboratory of Integrated Systems in September 2016. She is a program committee member of the International Conference on Advances in Circuits, Electronics and Micro-electronics, and has served as a reviewer for IEEE TCAD, IEEE TVLSI, IEEE TEVC, ACM JETC, ACM TOCL, Microelectronics Reliability, and MVLSC, as well as many conferences in the area of VLSI design and electronic design automation. Her research interests are currently focused on logic synthesis, in-memory computing, multi-objective optimization, and evolutionary computation.

Chapter 1
Information Processing and Storage in the Brain

Manfred Fahle

Abstract This chapter aims to illustrate the reception, encoding, and storage of information—including its modification through learning—in the brain of mammals, especially humans. The presentation of these processes begins on the level of a black-box analysis, that is, with a description of behavior, including patient studies that measure the capabilities and limitations of the entire system. The next level investigates the brain and its function based on the anatomy and histology especially of the human cortex revealing a functional specialization of cortex as already found by Brodmann (Vergleichende Lokalisationslehre der Grosshirnrinde. In ihren Priczipien dargestellt auf Grund des Zellenbaues. Barth, 1909). This functional specialization can also be seen in the various imaging studies based on the recording of the electric or magnetic brain activity or on changes in blood flow in the brain (functional magnetic resonance imaging). The third level is that of single cell recordings, mostly in animals, which shows the properties of single neurons and small neuronal assemblies. The last and most basic level considered is that of the biochemistry of information processing. On this level, description of the cellular mechanisms underlying information storage is most prominent. Modeling will help understand the experimental results on each of these levels.

1.1 Introduction

1.1.1 Background: Information Processing and Storage in the Brain

For clarity, we should begin by going back to the definition of the word "information" as outlined in the introduction to this book. Depending on the context, there are many different definitions of the word "information." For example, in

M. Fahle (✉)
Zentrum für Kognitionswissenschaften, University of Bremen, Bremen, Germany
e-mail: mfahle@uni-bremen.de

C. S. Große, R. Drechsler (eds.), *Information Storage*,
https://doi.org/10.1007/978-3-030-19262-4_1

psychology and perhaps also in everyday parlance, information may be defined as the ensemble of "*facts provided or learned about something or someone*" [20]. In computing, on the other hand, information may be defined as "*data that are processed, transmitted and/or stored by technical devices*" (loc. cit.). Information theory, finally, has an even wider definition of information, for example, "*as a mathematical quantity expressing the probability of occurrence of a particular sequence of symbols, impulses, etc. as against that of alternative sequences*" *(loc. cit.)*.

Here, I want to outline how information is processed and stored in the brain of primates, especially humans. How did we come to believe that information is mainly stored and processed in the brain? There were times in history when, for example, some Greek philosophers thought of the brain as a mere cooling mechanism with a large blood supply and a large surface that could produce a lot of sweat on the head's surface. But even then, some of the old Greek philosophers disagreed with this view and thought of the brain rather than the heart as the organ of thoughts and emotions. In the end, this view prevailed.

For a period of time, people thought of brain functions in terms of hydraulics. This was in the age of the steam engine and it is true that the brain receives a lot of blood supply. Later on, in the age of the emergence of fine watches, some people drew parallels between mechanical devices and brain function. There is, for example, a somewhat strange story by E.T.A. Hoffmann about a mechanical woman made from wood ("Olimpia" in "Der Sandmann"). The question as to how exactly the brain collected information about the world and how it learned received no particular attention, apart from visual perception, with some Greek philosophers such as Euclid (300 BC) suggesting that the eyes project light to the outer world that "senses" the world in a way similar to what fingers do.

Now we know that the brain relies on electrical activity but is not a computer, an analogy that was popular with some people about 50 years ago. Today, we feel safe to say that the brain consists of a number of adaptive, recurrent networks that are associative and rely on electric activity. These networks, for example, enable humans to store different types of information on different time scales, and the storage time of information is one way of classify information storage in the brain. Another way to classify information processing is by discriminating between information that can be coded in words, the so-called declarative information about facts and events (last Christmas, the French revolution), and non-declarative information, like in procedural skills (riding a bike, playing piano) or associative or operant conditioning (cf. Pavlov's dog).

1.1.2 Some Problems That Have to Be Solved: From Outside Objects to Meaningful Internal Representations

Already the ancient Greeks raised the question how information about the outer world can enter the brain. The atomists resolved the problem by postulating that all objects send out miniature/atomic copies of themselves that enter the eye and from there reach the brain. But already Socrates (470–399 BC) proved this view to be wrong, for example, by pointing out that objects outside the central visual field are only rudimentarily perceived. In this chapter, we will outline our contemporary views on how information enters the nervous system through specialized peripheral receptor organs, is encoded by electric activity in nerve cells and transmitted to the central nervous system. Similarly well understood are the cellular and biochemical mechanisms that underlie the storage of information in the central nervous system. This part of information processing is indeed very well if not completely understood.

But there are other questions that are even more difficult to answer. For example, how does the brain decide between important and unimportant information? The information content in bits per second supplied by the optic nerve alone is far higher than what the human brain can process. Hence, the so-called attention selects the portion of information that seems to be most important. But what are the mechanisms of attention? And moreover, selection by attention alone would never be able to do the trick. Already on a rather low level, there are mechanisms that compress the incoming information and remove redundancies. Attention, whatever this might be, employs knowledge in a top-down way to filter the incoming information, while at the same time being triggered by abrupt changes in the input from external sources. On the next level, the question arises as to which information to store and for how long. In order to make information about the outer world useful, it is necessary to compare inputs with stored templates, hence to generalize and to categorize inputs. An important question is: how does the brain categorize and generalize? How can it infer the future from the past?

For some types of information brains are able to process many parallel sources of inputs, while other types of information have to be processed in a serial manner. Stimuli from the visual domain are good examples. We are able to analyze, at one glimpse, the orientation of line elements over the entire visual field, that is, we analyze this feature in parallel (see Fig. 1.1a). This we achieve by hundreds of thousands "detectors," each devoted to one visual field position. The same is true for color (Fig. 1.1b). This extraction of contours and their orientation removes quite a bit of redundant information, a bit like a jpg reduces the storage required for an image—we reduce images a bit towards 'cartoons'. However, for small differences in orientation (Fig. 1.1d), for a combination of color and orientation (the oblique green stimulus in Fig. 1.1c), for a combination of features (vertical green and horizontal red; Fig. 1.1e) and conjunction (Fig. 1.1f) "the odd man out" has to be searched for, by means of a serial process that is governed by visual attention that "binds" together the different features of each element (e.g., color and orientation). It is also clear that a fair amount of information is processed subconsciously or

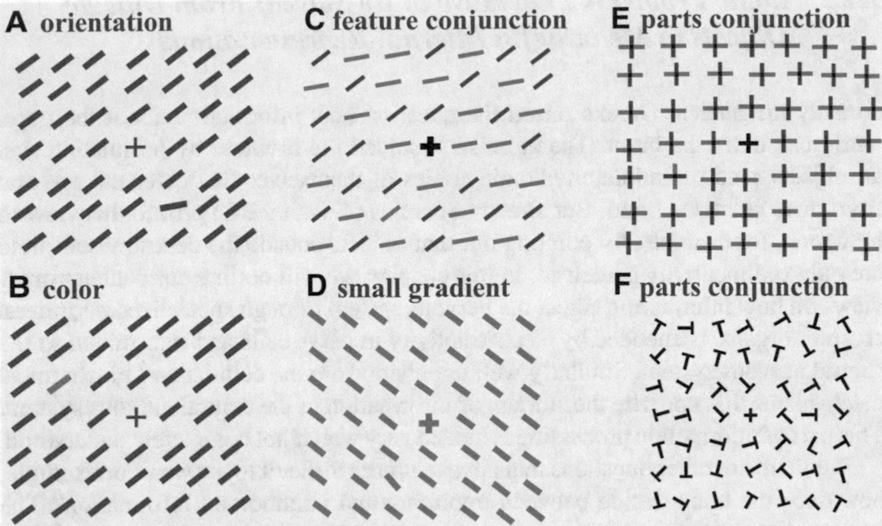

Fig. 1.1 (**a**) A red line stimulus among green stimuli (distractors) pops out, that is, it is immediately found, irrespective of the number of distractors (parallel search). (**b**) A target defined by color (e.g., red) *and* orientation (e.g., vertical) has to be searched for among red vertical and green horizontal distractors (serial search). (Figure reproduced from [14] Hochstein & Ahissar (2002): View from the Top Hierarchies and Reverse Hierarchies in the Visual System. Neuron 36, p. 793, with permission from © Elsevier, 2018. All Rights Reserved)

completely unconsciously. This is certainly true for most information regarding posture and walking. But in contrast to what you may read in the newspapers, the exact interactions within the neuronal networks underlying these different types of information processing, from attention over object recognition, space perception, and language-related processes to decision making and motor planning are far from clear.

We will look at the hardware, or should we say wetware, that stores information in the brain. Interesting questions here concern not only the mechanisms and mechanics of learning but also the many different types of data storage in the brain, for example, regarding the difference between short-term and long-term storage of information, the role of forgetting, and regarding the types of information that have to be stored since there is a big difference between information about objects, on the one hand, and on space, on the other hand. (We want to recognize objects irrespective of their position in space, hence generalize over space, while for interacting with objects, space information is crucial.) Moreover, one has to discriminate between data that can be stored and described by language versus other "knowledge" that cannot be expressed in words. In the spatial domain, there are important differences between storage in self-centered coordinate systems versus object-centered coordinate systems and it seems also between the near and the far space. A highly important point we will address is how the system is able to improve performance, usually called learning.

In my view, it is always important to scrutinize research results to find out by which methods these results were obtained. When somebody tells me my future, I am highly interested to learn how this person knows and came to the knowledge about my future. If this person tells me that my future can be read out of the positions of the stars, I personally am skeptical that his forecast will come true. Hence, we will look at the methods that brain research employs these days to study brain function in primates and especially humans.

1.1.3 Methods to Investigate the Brain: From Behavior and Patient Studies to Anatomy, Histology, EEG, and Single Cells

As outlined above, I believe that all results can only be as good as the method that has been used to reveal them. Therefore, I will not only present results on the processing and storage of information in the brain, but also shortly indicate the main methods employed to gain these results. However, for space reasons, a short overview has to be sufficient. For more details the reader is referred to textbooks such as the excellent one by [16].

Probably the earliest method to infer the processing of information and its storage in brains was to observe behavior, both in animals and in humans. Even in everyday life, we tend to infer from the behavior of others whether or not information has been collected and stored by these individuals. Elaborate methods exist to examine behavior with high precision both of animals and humans. For animals, especially operant conditioning—where one trains an animal, for example, to pull a lever in response to a stimulus—and associative conditioning are used—as with Pavlov's dog that learned to associate the ring of a bell with food. The methods employed with humans range from psychophysics (the quantification of sensory impressions) to personality psychology (classifying personality traits) and even psychoanalysis that deals also with the subconscious level of information processing. These investigations into normal brain function are complemented by studies in patients where deficits in function can be correlated with defects in brain structure and by modeling that helps to understand complex experimental results. This so-called black-box analysis (the name refers to the fact that all we know are the inputs into the system and its outputs, but not what happens in between, i.e., in the brain) is supplemented by methods able to give us some hints about what is actually going on in this black box. A relatively well-established example of these methods is the recording of EEG-potentials (electroencephalography). These can be recorded from the skull and reflect the activity of large ensembles of neuronal cells. The EEG-recordings are complemented by the recording of magnetic field potentials, MEG. A relatively new method relies on magnetic resonance imaging, called functional MRI, which is able to track the changes in blood flow in an intact brain when defined stimuli are applied.

On the next level are recordings of field potentials and/or single cells, usually in animals, but also in cell cultures and in rare instances even in humans. Additional new methods are able to visually record the activity of large portions of the cortex by applying either specific dyes that change their color depending on the activity level of the underlying cortex or by genetically modifying animals so that defined cell types in their brains can emit light when activated. On an even finer scale of observation, it is possible to study and to modify the transmitters by which nerve cells communicate with others—the field of neuropharmacology.

The methods used to study the *structure* of brains complement the *functional methods* outlined above: anatomy and histology that study not only the microscopic structure of brains but also their fine structure on the level of single cells (see Fig. 1.2). Additional methods, such as electron-microscopy and biochemistry, finally investigate both function and structure on the level below single cells, that is, on the level of communication between cells and how information in nervous systems is stored by modifications in the way in which cells communicate with each other and what the underlying biochemical and intracellular processes might be.

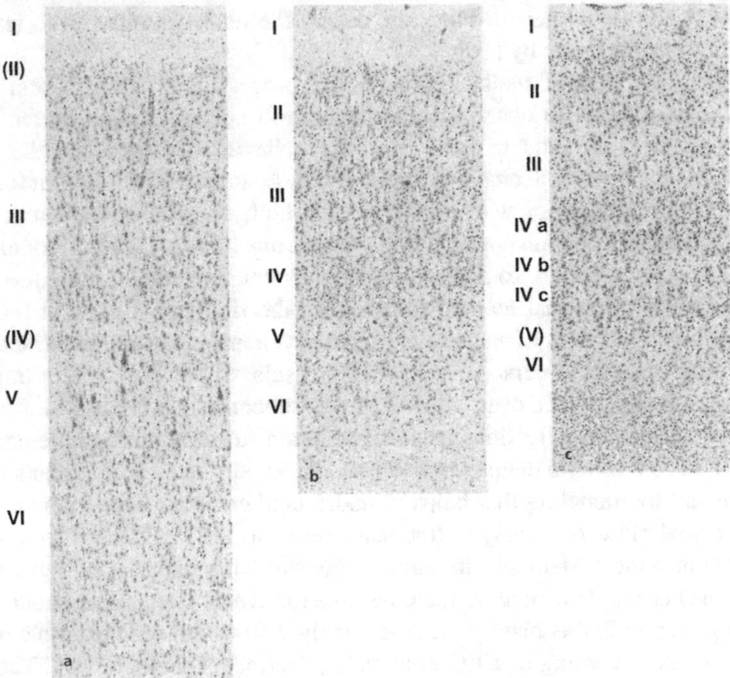

Fig. 1.2 Histological examples from different parts of the human cortex, clearly demonstrating a structural specialization that mirrors the functional specialization of cortex. As can be seen, the thickness of cortex varies between different parts of cortex, as does the pattern of distribution of neurons. On the basis of such differences, Brodman segmented the human cortex. The result is shown in Fig. 1.13. (Figure reproduced from [6, 7] Brodal (1969): Neurological Anatomy. In Relation to Clinical Medicine, p. 475, by permission of © Oxford University Press, 2018. All Rights Reserved)

It should have become clear from this short overview that the study of information processing and storage in brains is a highly interdisciplinary endeavor, encompassing disciplines such as biophysics, theoretical physics, biochemistry, biology, neuroinformatics, psychology, and even philosophy.

1.2 Behavioral Level

1.2.1 Collecting and Selecting Information About the Outer World

We will look at questions such as the ones outlined above on different levels, according to the methods used to obtain the results as indicated earlier. Let us start with the level of the black box or of phenomenology; that is, we will look at what can be inferred from investigating the relation between inputs to the brain and the outputs produced by the brain, be it by generating verbal responses or other types of behavior. It is customary to divide the hypothetical stages within the black box a bit according to introspection and on the basis of model assumptions. It sounds reasonable to postulate, as a first step, a conversion of an external stimulus by a sense organ into the type of information used by the brain. Obviously, sense organs are able to register a number of quite different carriers of information, for example, mechanical pressure on the skin, odor molecules streaming through the air from certain parts of our surroundings, as well as electromagnetic waves of a certain range of wavelengths ("colors"), and changes in air pressure that are quite low (usually called sounds). (We will not consider here information about the internal state of the body—this is a field by itself.)

The information collected by the sense organs has to be transmitted to the brain via nerve fibers, something we know from anatomy, but that can be inferred clearly even from black-box analysis: there must be some type of connection between the sense organs and the brain. We can also infer, from a black-box analysis, that there is a constant selection of information. The most clear external sign of this selection of information is human eye movements. We can measure the decrease of resolution from the center of gaze toward the periphery and find that already at 10° eccentricity, visual resolution is only about 20% of what it is at the center of gaze (Fig. 1.3). Clearly, that is the reason why we as humans move our eyes approximately once per second to gaze at the most "interesting" parts of a visual scene, guided by attentional processes. This change of gaze direction implies a constant selection of visual information in as far as only the objects at the center of gaze are resolved in any detail, while all the rest is not clearly perceived.

Through behavioral experiments we can also find out some of the criteria that govern the selection of information, for example, by testing which information that is supplied during an experiment has been stored and can be retrieved. We can also, through behavioral experiments, find out which types of features can be extracted

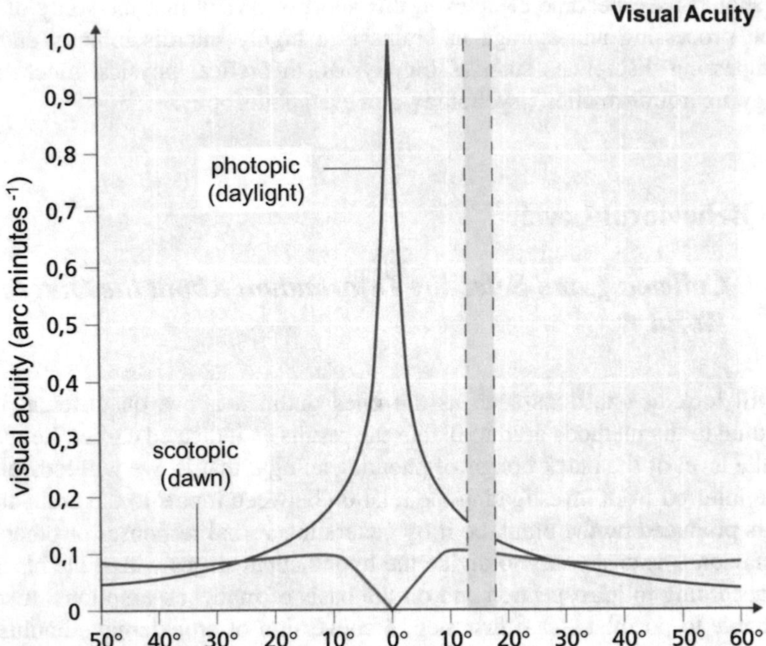

Fig. 1.3 Visual resolution of humans as a function of distance from the center of gaze. The dotted line indicates the effect of the blind spot, that is, the start of the optic nerve. Since only cones reside at the center of the retina and these do not function in the dark, acuity is best at about 10° eccentricity—where the maximum resolution by rods exists—during the night. (Figure adapted from [10] Fahle (2003): Gesichtssinn und Okulomotorik. Schmidt, R.F. & Unsicker, K. (Eds.) Lehrbuch Vorklinik Teil B: Anatomie, Biochemie und Physiologie des Nervensystems, der Sinnesorgane und des Bewegungsapparats. Kommunikation, Wahrnehmung und Handlung, p. 273)

by the brain in parallel, that is, simultaneously, for a number of criteria or for a number of frequencies or positions in outer space. A clear example is shown in Fig. 1.1 from [14]. A stimulus defined by a difference in color relative to that of its neighbors can easily be detected. But it is quite a difficult task to find a stimulus that is defined by a combination of a specific color and a specific orientation, when there are stimuli around that have either the same orientation (but a different color) or the same color (but a different orientation). For the latter task, we need time since we have to analyze if not each single item but at least groups of items sequentially by focused attention before we find the one stimulus that is unique.

Another example is from my own work. Finding a Vernier stimulus amidst stimuli that are aligned, as in Fig. 1.4a1, b1, is quite easy, while it is time consuming to find the Vernier that is offset to the right among Verniers offset to the left (Fig. 1.4a2, b2). So we can conclude that we do have detectors in the nervous system that can analyze the entire visual field regarding color, line orientation, and collinearity—enabling parallel, that is, simultaneous processing—while we need attention to "bind" color and orientation together as is indicated in the expression

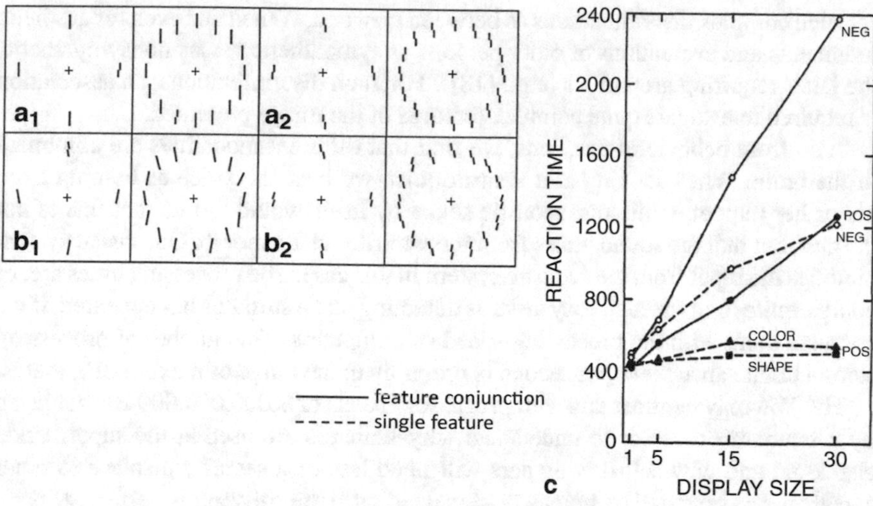

Fig. 1.4 Left (**a1** and **b1**) A target that has a small lateral offset, such as a Vernier target, pops out among distractors that are aligned. Left (**a2** and **b2**) A target offset to the left has to be searched among distractors that are offset to the right. (Figure reproduced from [9] Fahle (1991): Parallel perception of vernier offsets, curvature, and chevrons in humans. *Vision Research 31*, p. 2151 (top), with permission from © Elsevier, 2018. All Rights Reserved) Right (**c**) Time required to find a target increases linearly with the number of distractors for serial search, while not to find a target defined by a difference in color or in orientation (shape). *POS* target present in display, *NEG* target not present in display. (Figure reproduced from [26] Treisman & Gelade (1980): A Feature-Integration Theory of Attention. *Cognitive Psychology 12*, p. 104, with permission from © Elsevier, 2018. All Rights Reserved)

"spotlight of attention" and to discriminate between an offset to the left versus an offset to the right (serial processing). Figure 1.4c shows the increase of reaction times as a function of stimuli presented simultaneously. This is an important difference between the brain and standard computers that usually have a rather limited number of processors. The use of graphic cards for some computational tasks somewhat decreases this difference between brain and computers. Computers start to have many information-processing units in parallel, as has the brain, up to the point of in-memory computing (see Chap. 3).

1.2.2 Processing Times in Brains Versus Computers

We can also measure how long it takes before observers detect and/or identify a stimulus, for example, a visual stimulus. Detecting a stimulus requires less time (around 280 ms) than to categorize a stimulus (above 300 ms) [3]. Still, the identification and categorization task can be quite fast, as we just saw; it may not take much longer than the mere detection and reaction to a stimulus. This is even true for such complex discriminations as between men and women and even for aesthetic judgments and evaluations of other persons as sympathetic versus nonsympathetic, the latter requiring around 1 s (e.g., [18]). For such discriminations, an association is required to evaluate quite complex features of the image presented.

Also from behavioral reactions, we find that different modalities are combined in the brain. When looking at a ventriloquist, we hear the voice as coming from his or her puppet, while our acoustic sense by itself would tell us that this is not the case but that the sound arises from the ventriloquist's mouth. Our visual system modifies the input from the acoustic system in this case. The processing times are, of course, quite long for such easy tasks as detecting that a stimulus has appeared, if we compare them with the processing speed of computers. The number of processing steps a brain can achieve per second is probably in the range of maximal 10, that is, 10 Hz. You may contrast this with processor speeds of 2,000,000,000 Hz, achieved by a standard processor to understand why scanners are used at the supermarket check-out and why self-driving cars will need less of a security distance to other cars than cars operated by human drivers need once the software is perfected.

1.2.3 Information Storage in the Brain as Studied Behaviorally

As already mentioned, information can be stored in the brain in different ways, if we consider a black-box analysis leading to declarative versus nonverbal types of storage, and trying to retrieve stored information does not always succeed, as we all know from introspection. A synopsis of the different types of declarative versus non-declarative memory comes from the textbook by [15] (Fig. 1.5). It discriminates between clearly different types of storage that can be inferred from a purely behavioral standpoint and moreover indicates in the lower part of the figure brain structures in a somewhat tentative way that might underlie the storage of these different types of information.

As mentioned above, a second systematics of information storage on the behavioral level is the time interval over which information can be recalled. There is one type of storage that is extremely short, and that can be used, for example, to store phone numbers until we have something to write them down. This is called the articulatory loop, and as soon as we have to reply to somebody, to a question, we have lost the information about the phone number. Another extremely short-term memory, that is information storage, is the so-called iconic scratch-pad (photographic memory). It allows most of us to shortly memorize an image and is usually tested in patients by having them copy the figure displayed in Fig. 1.6, then to remove the figure and to have them draw the figure from heart (of course, it should be called by brain rather than by heart). On the medium-long time level, we have the so-called short-term memory which may last for hours or days and can be erased, for example, by concussion of the brain or by the application of electro-convulsion-therapy. And there is finally long-term memory that seems to be most stable and can last for a lifetime. We will come back to the mechanics underlying

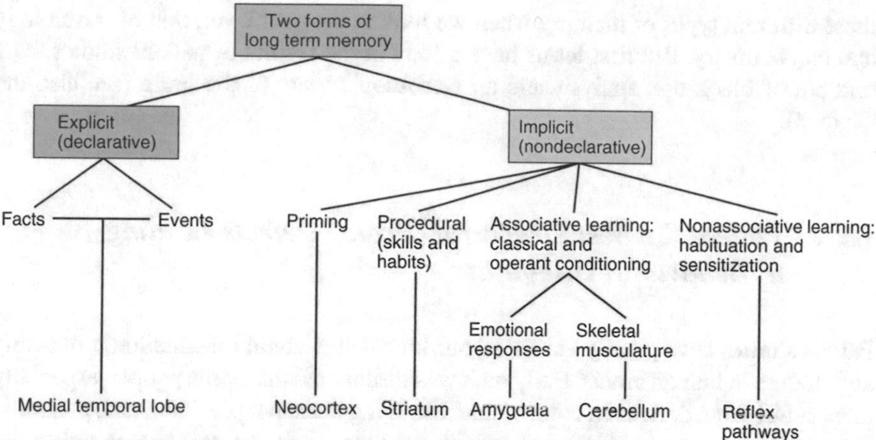

Fig. 1.5 A relatively large number of different types of learning and hence information storage exist in the human brain. Some information can be expressed in words (declarative), a large amount of learned content, however, is averbal (non-declarative). This textbook diagram presents also some ideas on where in the brain these different types of information are stored. (Figure reproduced from [15] Kandel et al. (2000): Principles of Neural Science, p. 1231, with permission from © McGraw Hill Companies, 2018 All Rights Reserved)

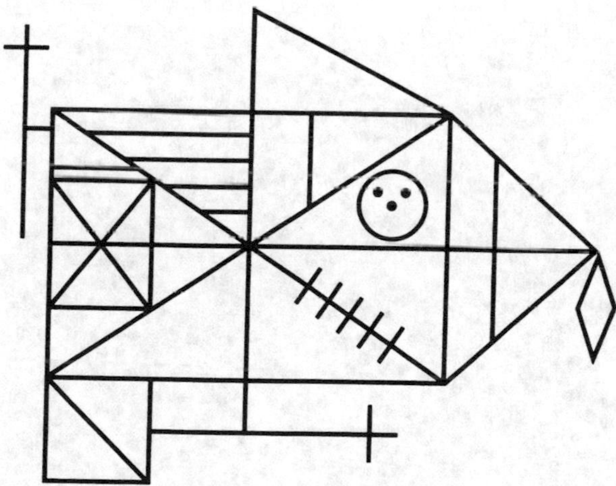

Fig. 1.6 The Rey-Osterrieth-Complex Figure. Subjects are asked to copy the figure while it is presented to them. Immediately after withdrawing the original (and in some cases additionally after a defined delay), subjects have to draw the figure by heart. The more elements they correctly draw, the higher is their score. (Figure adapted from: [23] Rey, A. (1941): L'examen psychologique dans les cas d'encéphalopathie traumatique. (Les problems.) Archives de Psychologie, 28)

these different types of memory when we look at the next level, that of physiology and biochemistry. But first let us have a look at the results of patient studies, as a mixture of black-box analysis and a macroscopic view of the brain (see also the Chap. 2).

1.2.4 Patient Studies: Structural (Brain) Defects Leading to Behavioral Disabilities

Patient studies have greatly enhanced our knowledge about information processing and storage in human brains. Early on it was discovered that some people, especially men, suffer from deficits in color vision. Studying these people, it became clear that there are at least three different photoreceptor types in the normal human retina and that each of these photoreceptor types can be disturbed selectively (see Fig. 1.7).

Another early insight was that the part of the brain that understands language is separated from the part that produces speech. It became even clear that after strokes, patients may not only suffer from motor disturbances or blindness, but also from more subtle disturbances of information processing, for example, from the inability to discriminate colors, to perceive motion, to judge distances, or to discriminate

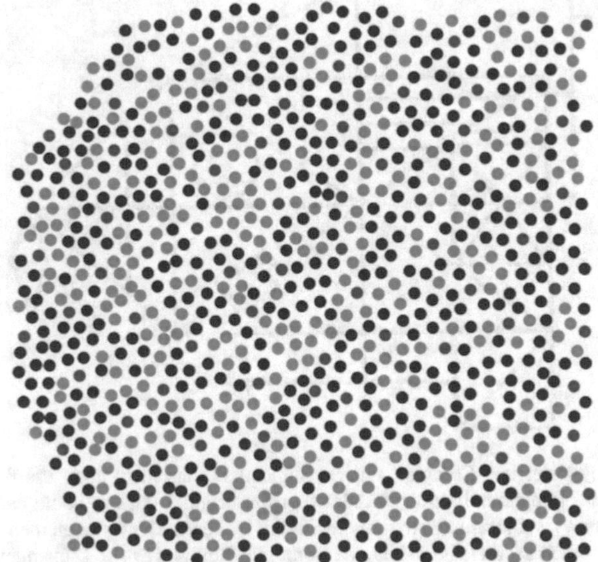

Fig. 1.7 The mosaic of photoreceptor-types in the human retina: Long-wavelength (L, red), medium-wavelength (M, green), and short-wavelength (S, blue) cones are shown as red, green, and blue dots, respectively. (Figure reproduced from [24] Roorda et al. (2001): Packing arrangement of the three cone classes in primate retina. Vision Research 41, p. 1295 (part d), with permission from © Elsevier, 2018. All Rights Reserved)

Fig. 1.8 Right part: Apperceptive agnosia. Patients suffering from this rare disorder can more or less clearly see individual dots and lines but are unable to combine these line elements perceptually to forms, let alone objects. Hence, patients suffering from an apperceptive agnosia are unable to copy even simple line drawings. Left part: Patients suffering from associative agnosia are able to copy line drawings but are unable to identify the objects they copied. (Figure reproduced from [1] Álvarez & Masjuan (2016): Visual agnosia. Rev. Clin. Esp. 216(2), p. 87; with permission from © Elsevier, 2018. All Rights Reserved)

between faces. From postmortem analysis, researchers were even able to localize the relevant parts of the cortex whose damage caused the symptoms mentioned.

Some patients may have normal visual resolution and normal low-level visual abilities, but suffer from disturbances in more complex visual functions [12]. One example is mentioned above, the inability to discriminate between faces. A more general disturbance is the so-called visual apperceptive agnosia. Patients suffering from this deficit have normal visual resolution and normal visual fields, but are unable to copy even simple line drawings, as is illustrated in Fig. 1.8. Other patients, when asked which objects they just copied, are unable to give an answer. These patients suffering from associative agnosia are also impaired at visually recognizing everyday objects while they might be able to name the same objects when these emit a typical sound. Hence, one can discriminate between patients who are unable to name an object that is visually presented, due to the fact that they cannot correlate the visual impression with their visual memory, that is, with the storage of objects that have been previously seen, on the one hand, and other patients who may still be able to do so but lack a connection to the corresponding word and who may be able to signal the meaning or function of an object by pantomime, on the other hand. Taken together, these deficits give the basis for a model of the sequence of processing steps in the processing of visual input, a final result of years of learning visual inputs during childhood to the evolution of visual categories that are stored in the visual association areas of the brain and that are connected to words that can be produced when the corresponding object is presented visually or acoustically or even haptically. Figure 1.9 shows a synopsis of these stages for the visual system.

A final example are two patients described by [13]. One suffered from a pronounced deficit in object recognition similar to the one described in Fig. 1.8, but had no problems finding her way in space (her "what"-pathway was defective).

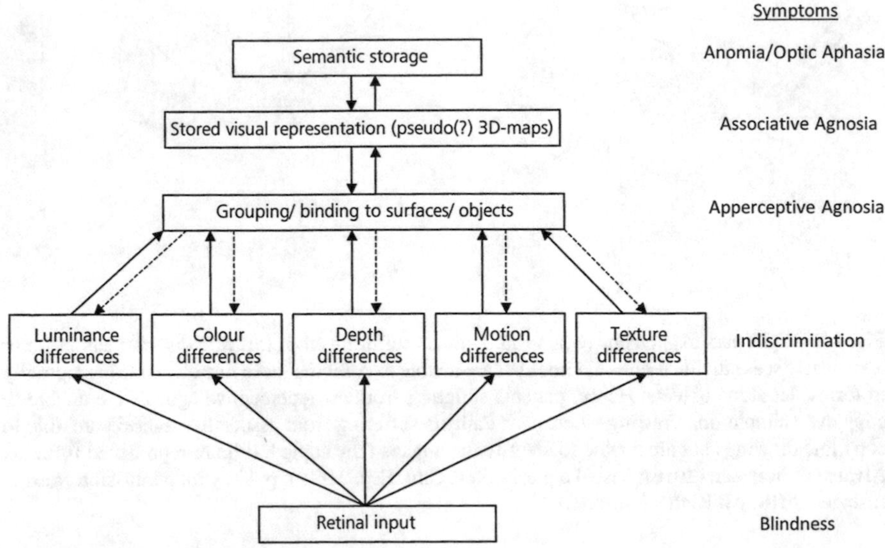

Fig. 1.9 Schematic overview of the levels of cortical information-processing and the symptoms arising from disturbances on the different levels. On the first level beyond the retina, boundaries are detected based on transitions in luminance, hue, (stereoscopic) depth, motion-direction or speed, texture, or other cues, for example, second-order features in these submodalities. Defects on this level lead to symptoms that are called indiscriminations. On the next level, contours or boundaries are combined and bound together to form objects. Defects on this level lead to apperceptive agnosias. On a third level, the objects are compared with stored representations of objects encountered earlier, that is, the present object is categorized and hence recognized. Defects on this level lead to associative agnosias, for example, to prosopagnosia if only recognition of faces is defective or alexia if the defect concerns words. The recognized object is usually linked to a noun (semantic storage). Missing of this link leads to optic aphasia. (Left figure and legend from [11] Fahle & Greenlee (2003): The Neuropsychology of Vision, p. 196; by permission of © Oxford University Press 2018. All Rights Reserved)

The other one, with a structural defect in another part of the brain, was able to easily identify objects but suffered from massive problems in moving through space and in identifying the orientation of objects in space (her defect was in the "where"-pathway; see Sect. 1.3.6 for a description of the "what" versus "where" cortical pathways).

Another insight from patient studies is that information that is stored in the brain might not be readily retrievable. We all know from everyday experience that sometimes we have problems to retrieving information that we know is somewhere in our brain. Patients with strokes of the brain often suffer, if temporary, from so-called neglect. These patients have only reduced or no access at all at information regarding usually the left side of their bodies and/or outer space. Well-known examples of symptoms produced by such patients are that a patient complains of not having enough food because he or she ate only the right part of the food present on the plate. Rotating the plate by 180° will satisfy such a patient.

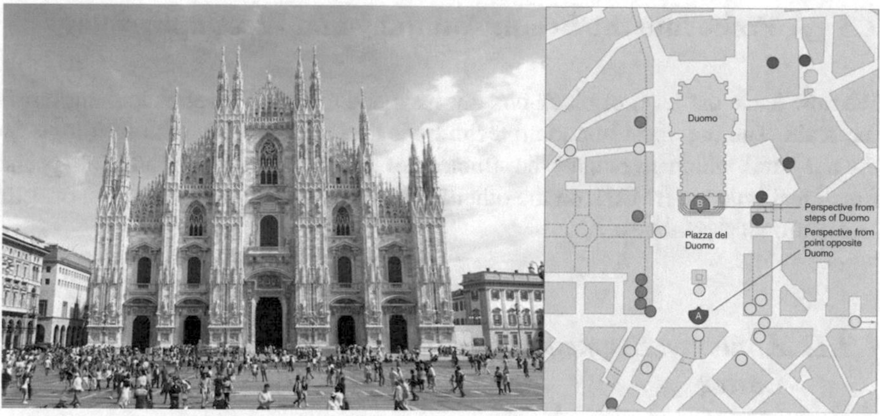

Fig. 1.10 An experiment by Italian neuropsychologist Edoardo Bisiach clearly shows that information stored in the brain cannot be accessed in these neglect patients. His patients remembered only part of the buildings at the central place in Milan, depending on where they imagined to be standing. (Left image from: https://pxhere.com/en/photo/744007. Right figure reproduced from [16] Kandel et al. (2013): Principles of Neural Science, p. 384, with permission from © McGraw Hill Companies, 2018 All Rights Reserved. Adapted from [4] Bisiachi & Luzzatti (1978): Unilateral neglect of representational space. Cortex 14, 129–133, with permission from © Elsevier, 2018 All Rights Reserved)

In a nice experiment, the Italian neuropsychologist Edoardo Bisiach [4] demonstrated that the information about outer space is still present in these patients. He asked patients suffering from neglect to imagine standing in front of the dome at the central place in Milan and to describe which buildings they would see (Fig. 1.10). Surprisingly, these patients mentioned only those buildings that were on the right side of their imagined visual field. Later on, he asked them to imagine standing on the other side of the same place, now looking towards the dome. Again, they were asked to report all the buildings they saw in their imaginary field of view, and again, they mentioned only the buildings on their imagined right side. This result shows that the information about all the buildings at this place was present in their brains but they could only access part of this information.

A somewhat related phenomenon in normal subjects is so-called priming. When a list of, say, 20 words is read aloud and people are then asked to write down as many words of this list as they can remember, the result is that, depending on the length of the list, for example, only half of the words might be written down. If, on the other hand, people are asked to complete letter sequences consisting of three letters (such as abs ...), then they tend to produce the words they had heard as part of the list that was read aloud (for example "absurd")—even those they did not consciously remember, hence "absurd" rather than "absent," "absolve," "absorb," "absinth." This effect is most pronounced in patients suffering from so-called amnesic syndrome who will actively remember only very few of the words but will achieve about the same number of hits as normal people, if the task is not to remember the words but to complete three-letter strings.

1.3 A Probe into the Brain: Anatomy and Electrophysiology

We now leave the level of black-box analysis and enter the level of looking inside the brain. The two most important techniques here are anatomy with histology, on the one hand, which investigate the structure of the brain, and electrophysiology and functional imaging (fMRI), on the other hand, which examine the function of cells and neuronal assemblies.

1.3.1 Anatomy: Segmentation and Specialization of Human Cortex

Let us start with anatomy. Figure 1.11 shows a lateral view of a human brain where different colors indicate different parts of the cerebral cortex. It can be seen that the posterior part of the brain is almost exclusively devoted to processing inputs, be they visual, acoustic, or somatosensory. The anterior part on the other hand is almost exclusively devoted to planning and producing motor acts.

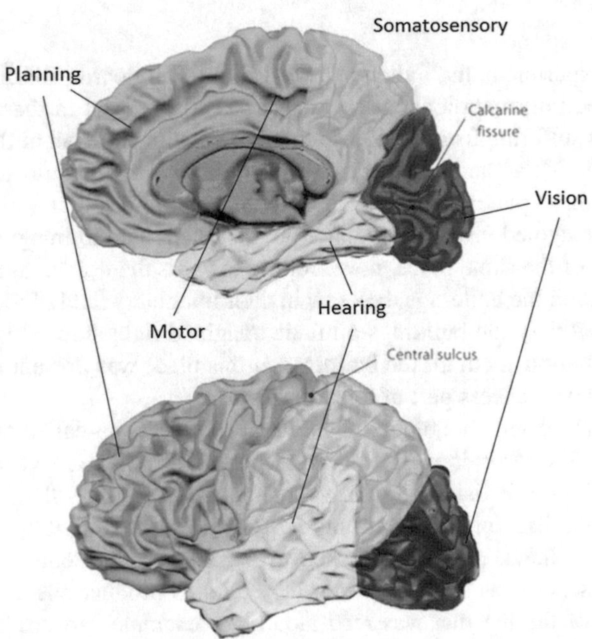

Fig. 1.11 Bottom: Lateral view of a human brain (that of the author) representing the surface between cortical grey and underlying white matter. Blue, frontal lobe; green, parietal lobe; yellow, temporal lobe; red, occipital lobe. Top: Medial view of the same brain. (Figure and legend after [11] Fahle & Greenlee (2003): The Neuropsychology of Vision (Plate 3), by permission of © Oxford University Press, 2018. All Rights Reserved)

1.3.2 *Histology: The Elements of Brains—Neurones*

The brain consists of about 10 billion neurones with up to 1000 connections to other neurones. Figure 1.2 shows all the cell bodies in thin slices of brain tissue while Fig. 1.12 shows about 1% of neurons including their arborizations (dentrites and axons) in a somewhat thicker slice. If all cells had been impregnated, the slice would be solid black. The great thing about these silver stains is that not only do they stain only a limited number of neurons but they moreover stain those neurons they choose in their entity, that is, with all their axomal and dentritic arborizations. The image in Fig. 1.12 gives a good impression of how the elements of information processing in the brain actually look like. Of course, there are quite a number of different types of neurons. But they have in common that they have first, an input part with so-called dendrites where they receive (mostly) chemical activations from other cells. Second, they have a cell boy with the cell nucleus and the genetic information and third an axon that may have many arborizations and up to 1,000 or even more connections to other cells. On the basis of differences in the fine structure of cortical architectures, as are evident in Fig. 1.2, Brodman [7] segmented the human brain into around 50 different areas. Only much later it became clear that these differences in fine structure are the basis of different functional specializations, such as seeing, hearing, or moving (see Fig. 1.13).

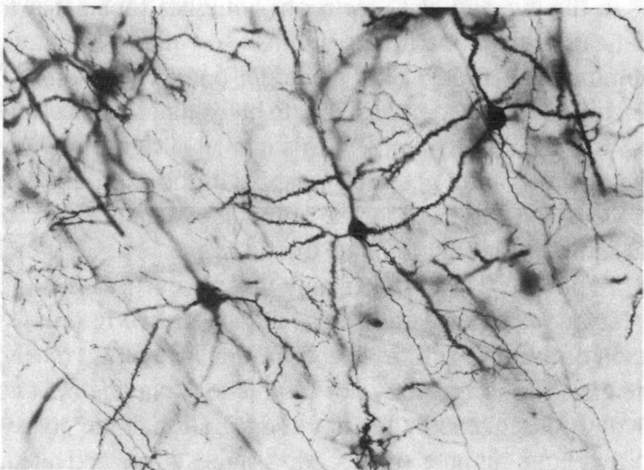

Fig. 1.12 The Golgi silver stain (named after its inventor, Camillo Golgi) stains only about 1% of neurons in a slice, but these neurons are stained completely, giving a clear impression of the structure of the cortical machinery. (Figure reproduced from [5] Braitenberg (1977): On the Texture of Brains. An Introduction to Neuroanatomy for the Cybernetically Minded, p. 107, with permission from © Springer Nature, 2018. All Rights Reserved)

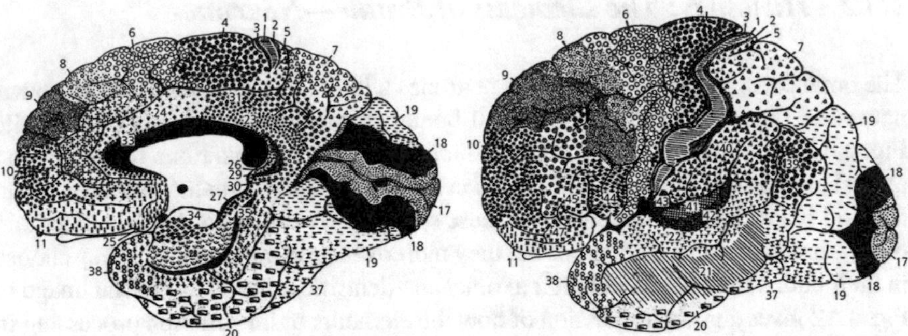

Fig. 1.13 Segmentation of the human cortex on the basis of cytoarchitecture (from [7] Brodman 1909). The numbers indicated were assigned by Brodman himself—unfortunately, his system of assigning numbers is difficult to understand. (Figure and legend after [11] Fahle & Greenlee (2003): The Neuropsychology of Vision, p. 185, by permission of © Oxford University Press 2018. All Rights Reserved)

1.3.3 Electrophysiology: From Generator Potentials to Orderly Cortical Representations

Electrophysiology can tell us some of the properties of these neurons. By recording from receptors in the eye and ear, electrophysiologists found that these receptors react to the adequate stimulation by producing a so-called generator-potential. The stronger the stimulus, the stronger the generator-potential becomes. This is to say that the transduction from an external signal to body-inherent information produces, as the first step, an analogue signal. But this analogue signal is transformed into a digital one, namely, the action potential or spike in the axon that leads to the next station toward the brain. In the axon, the stimulus' strength is no longer coded in an analogue fashion, but by the number of spikes per second, which all have the same form and amplitude. These spikes may then be integrated at the next synapse via smaller or larger amounts of transmitter that is set free into the synaptic cavity between the axon of the "transmitting" and the dendrite of the receiving neuron. A few action potentials per second may not be able to excite the next neuron and will vanish without any consequences, due to the integration in the postsynaptic neuron that may require a large number of incoming spikes to be activated, for example, more or less simultaneous inputs from several neurons.

In the acoustic system, there is a separation between stimuli on the basis of their frequency already in the inner ear, while in the visual system this separation is primarily on the basis of spatial position and, of course, on the basis of color or, more precisely, wavelength. From the receptor organs, signals are conducted to specialized parts of the cortex, as indicated in Fig. 1.14. The visual system, which

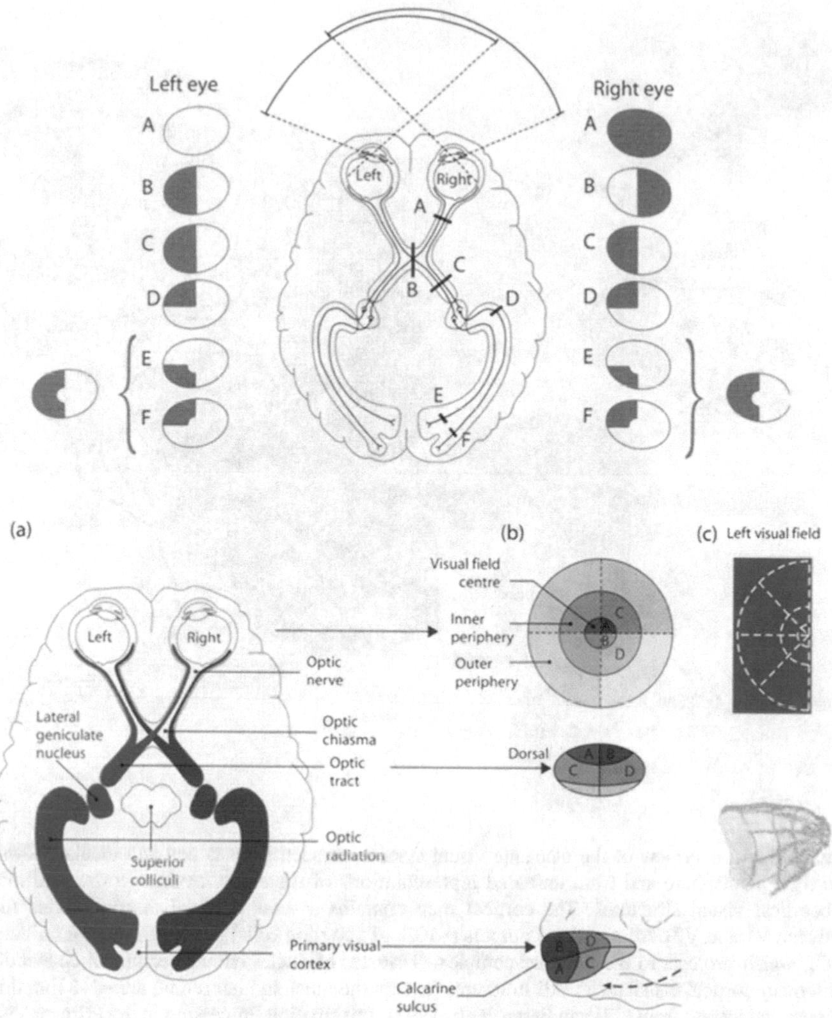

Fig. 1.14 Top: Visual field defects (scotoma) caused by lesions at different levels of the visual system. Retinal lesions produce unilateral, incongruent scotoma in the affected eyes (and hemifields), and so do optic nerve defects, (A) lesions of the chiasm, (B) optic tract, (C) lateral geniculate nucleus (LGN), and optic radiation (D, E, F) produce lesions only in the contralesional hemifield that are roughly congruent, that is, cover the same area when tested monocularly. The same is true for the primary, or striate, visual cortex, area 17. Lesions of extrastriate cortical areas often create blindness in one quadrant while the other one is spared, due to the fact that the representation of the upper and the lower visual field are separated in extrastriate cortical areas. (Figure and legend from [11] Fahle & Greenlee (2003): The Neuropsychology of Vision, (Plate 1), by permission of © Oxford University Press 2018. All Rights Reserved.) Bottom: (**a**) Projection from both halves of both retinae to the LGN and primary visual cortex. (**b, c**) Topography of the projection of the visual world on to the primary visual cortex. The center of gaze, corresponding to the fovea, is represented at the occipital pole. ((**c**), lower part). Visualization of the projection to the primary visual cortex. The stimulus displayed in the upper part of the graph produces the cortical activation displayed in the lower part. (Figure and legend from [11] Fahle & Greenlee (2003): The Neuropsychology of Vision, (Plate 2), by permission of © Oxford University Press 2018. All Rights Reserved)

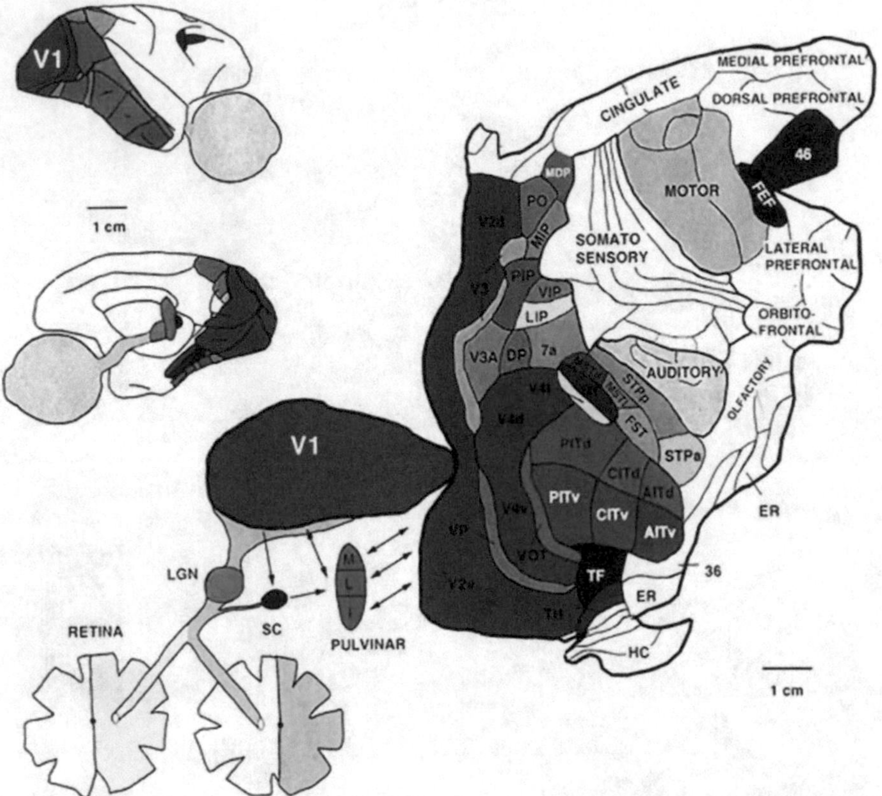

Fig. 1.15 An overview of the macaque visual system, as seen from lateral and medial views of the right hemisphere and from unfolded representations of the entire cerebral cortex and major subcortical visual structures. The cortical map contains several artificial discontinuities (e.g., between V1 and V2). Minor retinal outputs (~10% of ganglion cells) go to the superior colliculus (SC), which projects to the pulvinar complex, a cluster of nuclei having reciprocal connections with many cortical visual areas. All structures (except the much thinner retina) are ~1–3 mm thick. (Figure and legend from [27] van Essen et al. (1992): Information Processing in the Primate Visual System: An Integrated Perspective. Science, 255, p. 420, reprinted with permission from © AAAS, 2018. All Rights Reserved)

is certainly the best studied of the sensory systems, encompasses about 50 different cortical areas that can be distinguished from each other on the basis of histology and/or response-characteristics of their neurons. Figures 1.15 and 1.16 show that together these areas make up about a third of the macaque monkey's cortex. This means that the processing of visual information is very important for mammals and especially primates. In line with the results of patient studies, some of areas are specialized to analyze motion or color or depth or faces or places.

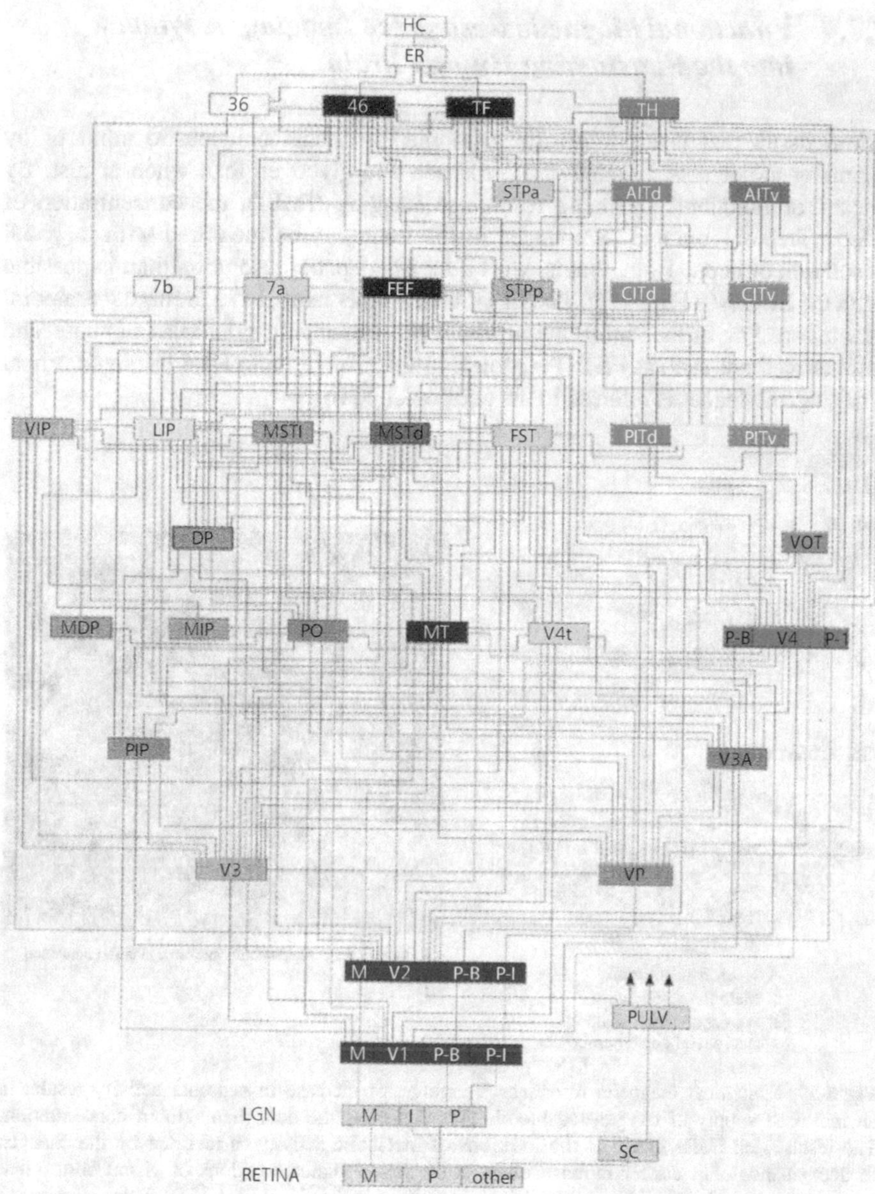

Fig. 1.16 Schematic view of the cortical areas involved in visual analysis in the monkey brain, based on the pattern of axonal connections between areas, as well as on differences in cytoarchitectonic and electrophysiological properties of single neurons. (Figure from [27] van Essen et al. (1992): Information Processing in the Primate Visual System: An Integrated Perspective. Science, 255, p. 421, reprinted with permission from © AAAS, 2018. All Rights Reserved)

1.3.4 *Functional Magnetic Resonance Imaging: A Window into the Functioning Human Brain*

Brain tissue that is activated, for example, by visual or acoustic input or by planning motor acts, requires more oxygen and glycogen than when at rest. By means of functional magnetic resonance imaging (fMRI), the concentration of (deoxy)hemoglobin and its changes in the brain can be measured with a spatial resolution of about 1mm, that is, with a far finer spatial resolution than is possible with the EEG (see Fig. 1.17). The EEG, on the other hand, has a far higher temporal resolution. For further information about the underlying principles of these and similar methods such as PET (Position Emitting Tomography) and infrared cortical imaging, the reader is referred to the textbook by [16].

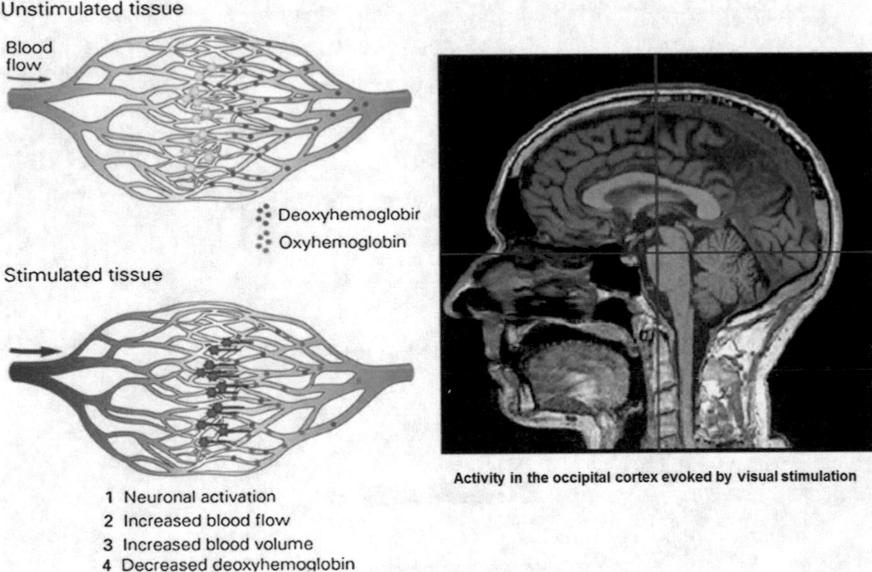

Fig. 1.17 Functional magnetic resonance imaging. An increase in neuronal activity results in an increased supply of oxygenated blood. This decreases the deoxyhemoglobin concentration. The result is an fMRI image of the locations of metabolic activity as revealed by the changes in deoxyhemoglobin concentration. Colors in the image indicate regions of visual cortex that responded to visual stimuli placed at particular locations in the visual field. (Left figure reproduced from [16] Kandel et al. (2013): Principles of Neural Science, p. 432, with permission from © McGraw Hill Companies, 2018 All Rights Reserved. Legend after Kandel et al., 2013, p. 432. Right figure from Stephanie Rosemann, with permission from © Rosemann, 2018. All rights reserved)

1.3.5 Imaging Studies on Top-Down Influences: Attention and Expectation

Electrophysiology has also shown that the properties of neurons can be changed not only by sensory input but also under top-down influences—which may be the mechanics of the so-called attention-processes. This is to say that the processing and filtering characteristics of even rather early cortical neurons can be modified under top-down influences, that is, under the influence of higher and more complex cortical areas that have stored information about the external world. These influences may represent the influence of learned contents and experience on early sensory information processing and lead us to expect certain inputs under some circumstances and others under different circumstances, for example, depending on whether we are at home or in a jungle to name two quite different circumstances. The results on perceptual learning as outlined below indicate that through learning—that is through modifications and storage of information—performance on these early levels of information processing can be modified. In a way, we are able to concentrate on specific features of sensory inputs depending on the task at hand.

1.3.6 Different Processing Pathways for Object Recognition Versus Space Processing as Defined by Electrophysiology and Imaging Studies

In order to interact with the outer world, the brain has to solve two quite different problems. The first is to identify visual objects, and the second one is to analyze the spatial structure of the surroundings in order to interact with objects, as mentioned above. To solve the first task, identification of objects, it is desirable to generalize across space so that you can identify an object irrespective of its location in space. The temporal or ventral pathway of visual information processing is specialized for this task. On the other hand, to navigate through space and to interact with objects of the outer world, the exact object properties are less important than their exact position. The parietal or dorsal pathway of the mammalian cortex (see Fig. 1.18) is specialized for this task.

1.3.7 Stimulating the Cortex

It is possible to stimulate the cortex and to elicit specific responses. For example, applying a strong, short magnetic impulse through the intact skull over the motor cortex will produce a movement in the body part represented underneath this part of the skull. Stimulation over the posterior (visual) part of cortex will produce short

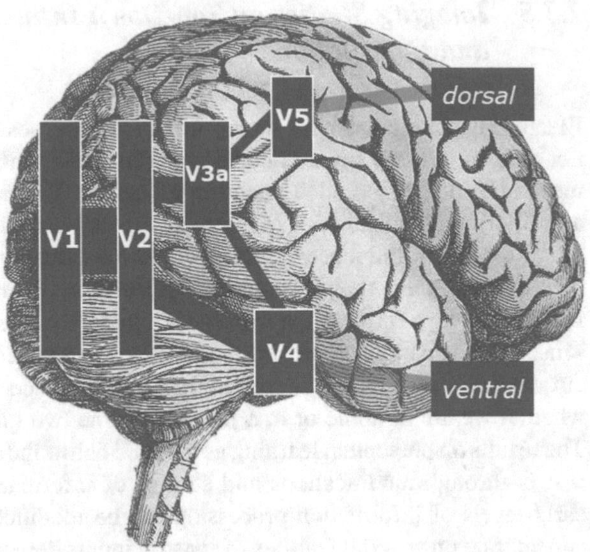

Fig. 1.18 The so-called ventral or "what" and dorsal or "where" pathways. Since the visual system has to solve two tasks: identifying objects irrespective of their location, and to exactly localize these same objects, two pathways of information processing evolved. The so-called dorsal pathway localizes objects (as a first step to interact with them), the ventral pathway identifies and categorizes these objects irrespective of their position in space. (Figure from Cathleen Grimsen, with permission from © Grimsen, 2018. All Rights Reserved)

flashes, called phosphenes. In patients, direct stimulation of cortex can even elicit memories of events and produce emotional responses [22].

In animals, recording from single cells can identify neurons whose discrimination of a stimulus moving slightly to the left versus to the right is more precise (if this response is fed into a computer) than that of the monkey [19]. This indicates that given the fact that a population of neurons is always responding to stimuli, the decision stage may not know which neuron is the best to follow, while the experimenter has carefully chosen one neuron with the (near) optimal signal-to-noise ratio. If now a neuron responding to, say, a movement to the right (or rather a small population of neurons) is stimulated electrically while a stimulus is actually moving slightly to the left, the monkey may signal that it perceived a movement to the right [25].

1.4 Single-Cell Studies in Animals and Humans and Biochemistry/Neuropharmacology

1.4.1 Elementary Properties of Single Neurons

Single-cell studies have greatly enhanced our knowledge regarding the properties of single neurons and small neuronal populations in the brain. Hence, we know that some cells in the primary visual cortex are specialized to detect (short) line segments of different orientations, with the entire visual field being represented by millions of neurons arranged in an orderly fashion (see Figs. 1.19 and 1.20). From the primary visual cortex (V1) with its six layers, information regarding color,

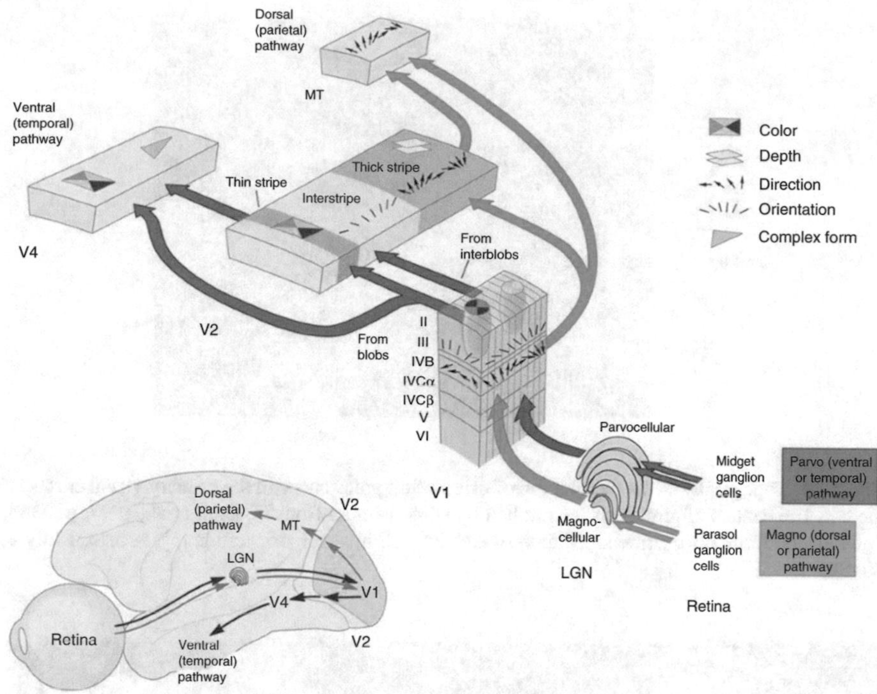

Fig. 1.19 Parallel processing in visual pathways. The ventral stream is primarily concerned with object identification, carrying information about form and color. The dorsal pathway is dedicated to visually guided movement, with cells selective for direction of movement. These pathways are not strictly segregated, however, and there is substantial interconnection between them even in the primary visual cortex. (LGN, lateral geniculate nucleus; MT, middle temporal area.) (Figure reproduced from [16] Kandel et al., 2013, p. 571, with permission by © McGraw Hill Companies, 2018. All Rights Reserved. Legend after Kandel et al., 2013, p. 571)

motion, orientation, depth, and complex form is relayed to different cortical areas and further analyzed separately in these areas, such as areas V2, V4, and MT.

Moreover, there are parts of the visual cortex that are specialized for the discrimination between different colors, different types of object classes, detection of faces, body parts, objects or places (see Fig. 1.21). In the auditory cortex, we find several tonotopic representations (see Fig. 1.22) and both the surface of the body and the muscles of the body are represented in an orderly fashion (see Fig. 1.23).

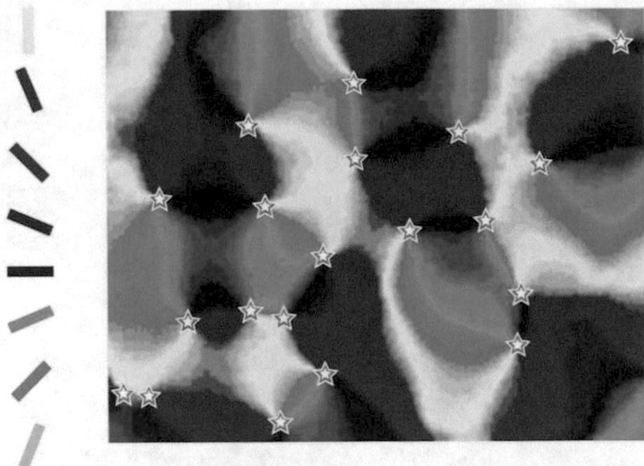

Fig. 1.20 Topography of the distribution of orientation preferences in the primary visual cortex of the cat. The centers of pinwheels are marked by stars. (Figure from [8] Crair et al., 1997, p. 3383, downloaded from: https://fr.wikipedia.org/wiki/Fichier:Pinwheel_orientation.jpg, accessed July 4, 2018)

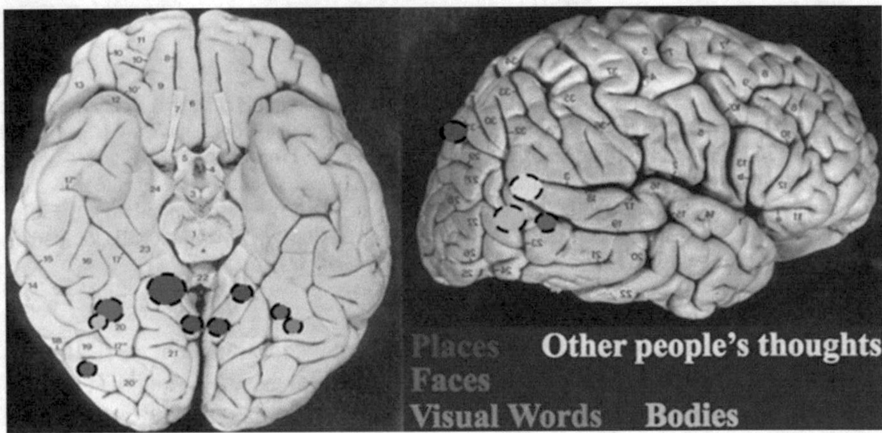

Fig. 1.21 This schematic diagram indicates the approximate size and location of regions in the human brain that are engaged specifically during perception of faces (blue), places (pink), bodies (green), and visually presented words (orange), as well as a region that is selectively engaged when thinking about another person's thoughts (yellow). Each of these regions can be found in a short functional scan in essentially all normal subjects. (Figure and legend reproduced from [17] Kanwisher (2010): Functional specificity in the human brain: A window into the functional architecture of the mind, PNAS 107 (25), p. 11164, with permission by PNAS, 2018)

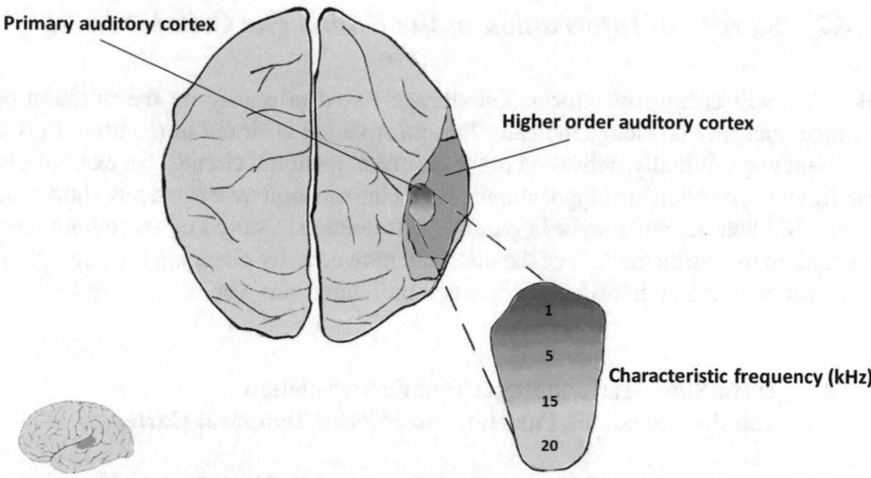

Fig. 1.22 The auditory cortex of primates is tonotopically organized. This means that there are stripes of cortex that are activated preferentially by the same frequencies and that these preferred frequencies change in a monotonic order over the cortical surface. The primary auditory cortex is surrounded by a number of higher-order areas that process no more single frequencies but more complex features of sound, up to areas devoted to understanding speech or analyzing music. Other parameters mirrored in the functional organization of the primary auditory cortex are response latency and loudness. The auditory map is only roughly present at birth (as is the case with the retinotopic map in the primary visual cortex) but evolves as a result of experience and learning

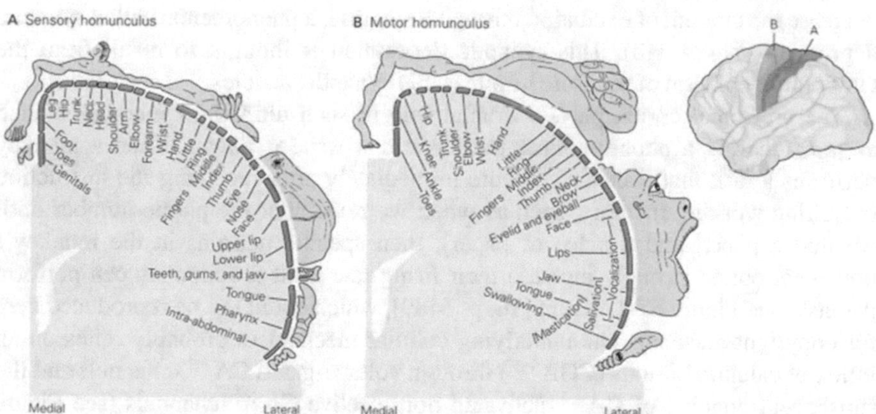

Fig. 1.23 A homunculus illustrates the relative amounts of cortical area dedicated to individual parts of the body. (Adapted, with permission, from [22] Penfield and Rasmussen 1950.) (**a**) The entire body surface is represented in an orderly array of somatosensory inputs in the cortex. The area of cortex dedicated to processing information from a particular part of the body is not proportional to the mass of the body part but instead reflects the density of sensory receptors in that part. (**b**) Output from the motor cortex is organized in a similar fashion. The amount of cortical surface dedicated to a part of the body is related to the degree of motor control of that part. (Figure reproduced from [16] Kandel et al. (2013): Principles of Neural Science, p. 364, with permission from © McGraw Hill Companies, 2018 All Rights Reserved. Legend after Kandel et al., 2013, p. 364)

1.4.2 Storage of Information in the Brain: The Cellular Level

Here we will categorize information storage based primarily on the duration of storage. Present knowledge indicates that information is stored in the brain in two different ways. Initially, activation in the recurrent neuronal circuits, for example, in the frontal cortex and the hippocampus, stores information over relatively short time intervals. Later on, and maybe in parallel, information is stored by more long-term changes in the wiring pattern of the neuronal networks, by changing the strength of information or signal transduction between individual nerve cells.

1.4.2.1 Ultra-Short-Term Storage: Tetanic Stimulation and the Prefrontal, Parietal, and Inferior Temporal Cortex

Let us start with the first type of information storage. Humans are able to store acoustic, visual, and abstract information over short time intervals, for example, by means of the phonological loop or the iconic scratchpad, as mentioned above. This short-time storage relies on changes in the excitability and firing patterns of (cortical) neurons. One prominent cellular mechanism underlying this change of excitation of synapses is an increase in the amount of transmitter-release. It has been shown that repetitive activation of a synapse produces a relatively persistent increase in transmitter release (see Fig. 1.24a) leading to post-tetanic potentiation. However, it should be noted that prolonged tetanic stimulation will, on the contrary, decrease the amount of excitation through a synapse, a phenomenon called synaptic depression (Fig. 1.24b). This synaptic depression is thought to result from the temporary depletion of the store of releasable synaptic vesicles.

The prefrontal cortex plays a crucial role in such ultra-short memories, such as remembering a phone number until you can write it down. When a monkey performs a task that it cannot execute immediately after receiving the instruction (requiring working memory, such as when we remember the phone number until we find a pencil and a piece of paper), then specific neurons in the monkey's prefrontal cortex strongly increase their firing rate until the monkey can perform the task (see Figure 67-1A from [16, p. 1489], which could not be reproduced here for copyright reasons). The underlying cellular mechanism probably relies on an influx of calcium^{++}-ions (CGA^{++}) through voltage-gated CA^{++}-channels and the subsequent opening of CA^{++}-activated nonselective CA^{++}-channels (see Figure 67-1 from [16, p. 1489], which could not be reproduced here for copyright reasons). Recurrent neuronal networks have been described of neurons in the prefrontal, parietal, and inferior temporal cortices that are synaptically connected and are able to create persistent reverberatory activity. Based on these findings, models have been developed to describe the function of such reverberating networks based on a combination of excitatory and inhibitory interactions between neurons (see Figure 67-1B from [16, p. 1489], which could not be reproduced here for copyright

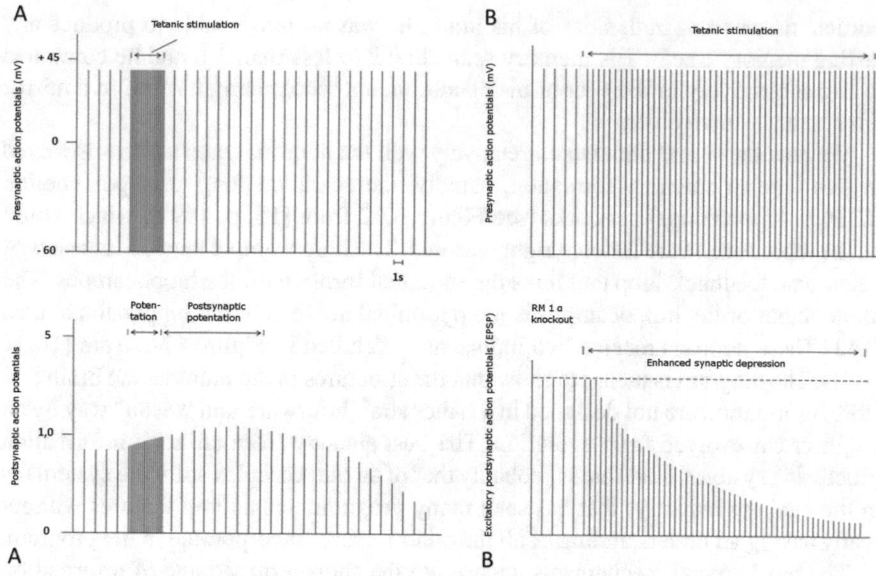

Fig. 1.24 A burst of presynaptic action potentials leads to changes in the resulting postsynaptic membrane potentials. (**a**) Upper part: A short but high-frequency stimulation leads to gradually increasing transmitter increase and hence to increasing postsynaptic membrane potentials, the effect called "potentiation." Note that the postsynaptic potentials remain enlarged even after the end of the presynaptic burst (lower part). (**b**) Upper part: A longer-lasting high-frequency stimulation may, however, decrease the amplitudes of the postsynaptic potentials (synaptic depression), probably due to a depletion of transmitter in the presynaptic membrane (lower part)

reasons). This type of memory can be extinguished, for example, by a concussion of the brain.

1.4.2.2 Medium-Term Information Storage: The Hippocampus

Information that has to be remembered for somewhat longer times is, to the best of our present knowledge, stored especially in the hippocampus. There, information can be stored for longer periods, ranging from days to years. To be more precise, it is not just the hippocampus but in addition a part of phylogenetically old cortex that lies near the hippocampus that plays an important role.

A first indication that this part of the brain stores information over somewhat extended times and is crucial for the transfer of short-term to long-term memory of explicit knowledge was obtained from patient Henry Molaison (HM). This patient underwent an operation in 1953 to cure epileptic seizures that could not otherwise be prevented. But after an operation that removed the hippocampus and adjacent cortical tissue from both sides of his brain, he was no longer able to produce any lasting memory traces. His memory span shrank to less than 1 h and he continued to live subjectively in the year of his operation, not recognizing himself in a mirror after he had grown older.

We now know and understand relatively well the neuronal mechanisms involved in this type of memory-formation, namely, the so-called long-term potentiation (LTP) in hippocampal networks (see Figure 67-2 from [16, p. 1491], which could not be reproduced here for copyright reasons). The hippocampal network consists of a neuronal feedback-loop that links the so-called fornix with the hippocampus. The main output of the hippocampus is the pyramidal neurons in the hippocampal area GA1. These neurons receive two inputs, as is detailed in Figure 67-2 from [16, p. 1491]. The diagram is meant to show that the structures in the mammalian brain that store information are not designed in a rather straightforward and "clean" way by an engineer but evolved from evolution. The best analogy from artificial information processing systems therefore is probably that of an old, complex software system (as in the banking industry) that has seen many programmers adding features without really having an understanding of all the other features incorporated in the program.

The biochemical mechanisms underlying the short-term storage of information (via long-term potentiation, LTP) are outlined in Fig. 1.25. Synapses involved in LTP can store information for durations between hours and weeks, maybe even longer. In these synapses, a strong, high-frequency stimulus through the input neuron (Fig. 1.25a, b) depolarizes the postsynaptic membrane more strongly than a single input as in Fig. 1.25a does. This leads to a de-blocking of the NMDA receptor, which is usually blocked by Mg^{++} ions, as can be seen by comparing Fig. 1.25a, b. Then, Ca^{++} ions enter the postsynaptic spine, resulting in four effects, as indicated (by arrows) in Fig. 1.25b. Again, it is not a simple straightforward effect but a combination of mechanisms on the cellular level leading to the storage of information. These effects can be complemented by still another one, the production, in the cell body, of additional receptors that are incorporated in the postsynaptic membrane (Fig. 1.25c).

As still another cellular mechanism to store information in the brain, LTP can activate so-called silent synapses (Fig. 1.26). Simultaneous electrophysiological (intracellular) stimulation in adjacent neurons in the hippocampus (see Fig. 1.26a) can activate the so-called silent synapses. Before LTP has been induced, neuron "a" cannot activate neuron "b," since the NMDA-receptor of neuron "b" is blocked (Fig. 1.26b, left). But when the membrane potential of neuron "b" is depolarized by means of the intracellular electrode, the postsynaptic membrane is excited by a spike coming from neuron "a" (Fig. 1.26b, right). After LTP, when the Mg^{++}-blockade of

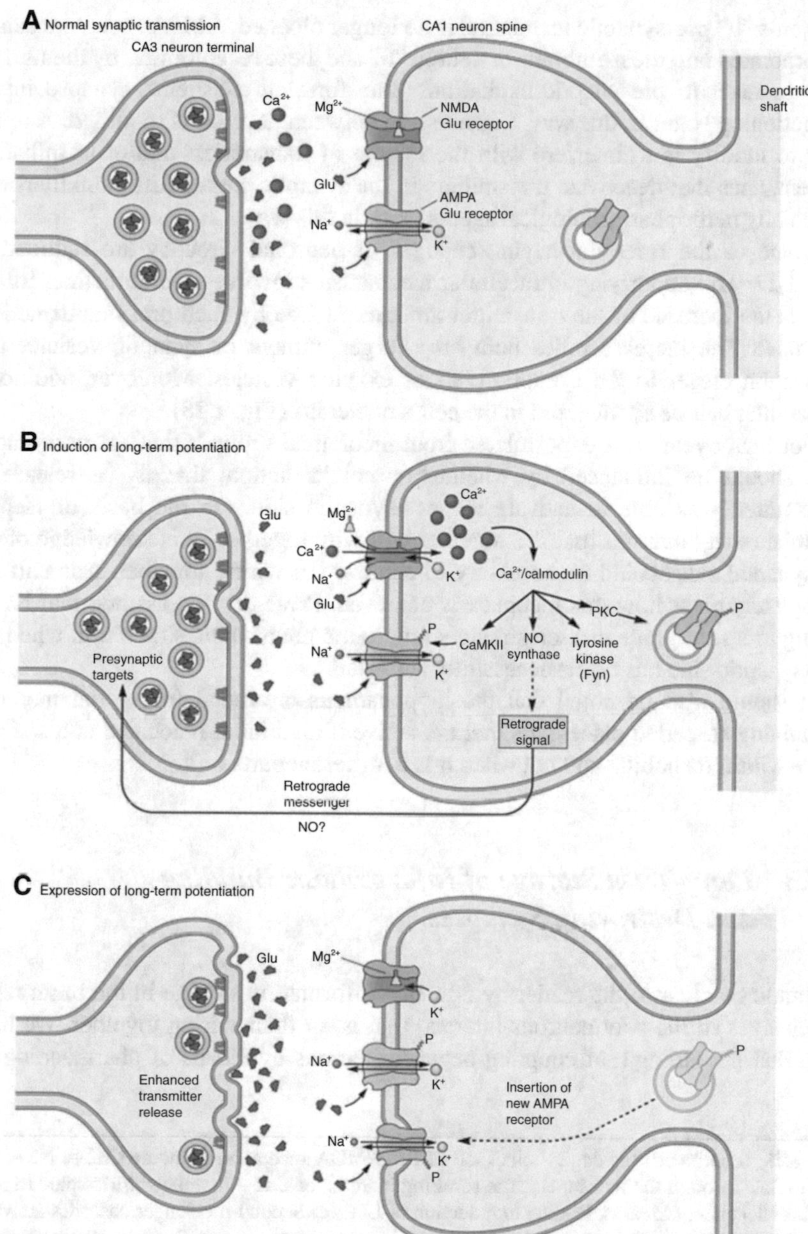

Fig. 1.25 A model for the induction of long-term potentiation. (**a**) During normal, low-frequency synaptic transmission, glutamate acts on both NMDA and AMPA receptors in the postsynaptic membrane of dendritic spines (the site of excitatory input). Sodium and K+ flow through the AMPA receptors but not through the NMDA receptors because their pore is blocked by Mg2+ at negative membrane potentials. (**b**) During a high-frequency tetanus the large depolarization of the postsynaptic membrane (caused by strong activation of the AMPA receptors) relieves

neuron's "b" postsynaptic membrane is no longer blocked, AMPA-receptors can be incorporated into the membrane of neuron "b" and these receptors are by themselves able to react to presynaptic excitations, transforming a "silent" synapse into a "functioning" one. In this way, transmission between neurons is modified. Another way to modify is to interfere with the release of transmitters and/or to influence the enzymes that deactivate transmitters in the synaptic cleft. Most painkillers and especially neuropharmacological agents work in this way.

Some of the rules underlying changes of neuronal circuitry are outlined in Fig. 1.27. The underlying intracellular mechanisms involve several distinct forms. One is the increase of the transmitter amount set free by each presynaptic action-potential. The increase relies both on a larger number of synaptic vesicles and a location closer to the synaptic cleft of existing vesicles. Moreover, additional transmitter can be synthesized in the cell's nucleolus (Fig. 1.28).

For many years it was postulated from theoretical grounds that the presynaptic side should be influenced by whether or not its action, that is, its release of transmitter, was able to activate the postsynaptic side. On the basis of Hepp's postulate that "neurons that fire together will wire together" this knowledge of the presynaptic side would be necessary to achieve the wiring together. It is still not completely clear how this feedback is achieved. Presently, we assume that NO is acting as a retrograde messenger, signaling to the presynaptic side if and when the postsynaptic side has been successfully activated.

It should also be noted that the hippocampus is crucial for spatial memory, containing so-called place-cells that are active if an animal is located at a specific place within its habitat and not when it is at different parts of it.

1.4.3 Long-Term Storage of Information: Building and Destroying Synapses

It should be clear to the reader by now that information storage in the brain relies on changes in the way neurons interact, that is, in their wiring together. We have seen that the strength of coupling between neurons by means of pharmacological

---

Fig. 1.25 (continued) the Mg2+ blockade of the NMDA receptors, allowing Ca2+, Na+, and K+ to flow through these channels. The resulting increase of Ca2+ in the dendritic spine triggers calcium-dependent kinases, leading to induction of LTP. (**c**) Second-messenger cascades activated during induction of LTP have two main effects on synaptic transmission. Phosphorylation through activation of protein kinases, including PKC, enhances current through the AMPA receptors, in part by causing insertion of new receptors into the spine synapses. In addition, the postsynaptic cell releases retrograde messengers that activate protein kinases in the presynaptic terminal to enhance subsequent transmitter release. One such retrograde messenger may be nitric oxide (NO), produced by the enzyme NO synthase (shown in part **b**). (Reproduced from [16] Kandel et al., 2013, p. 1494, with permission by © McGraw Hill Companies, 2018. All Rights Reserved. Legend after Kandel et al., 2013, p. 1495)

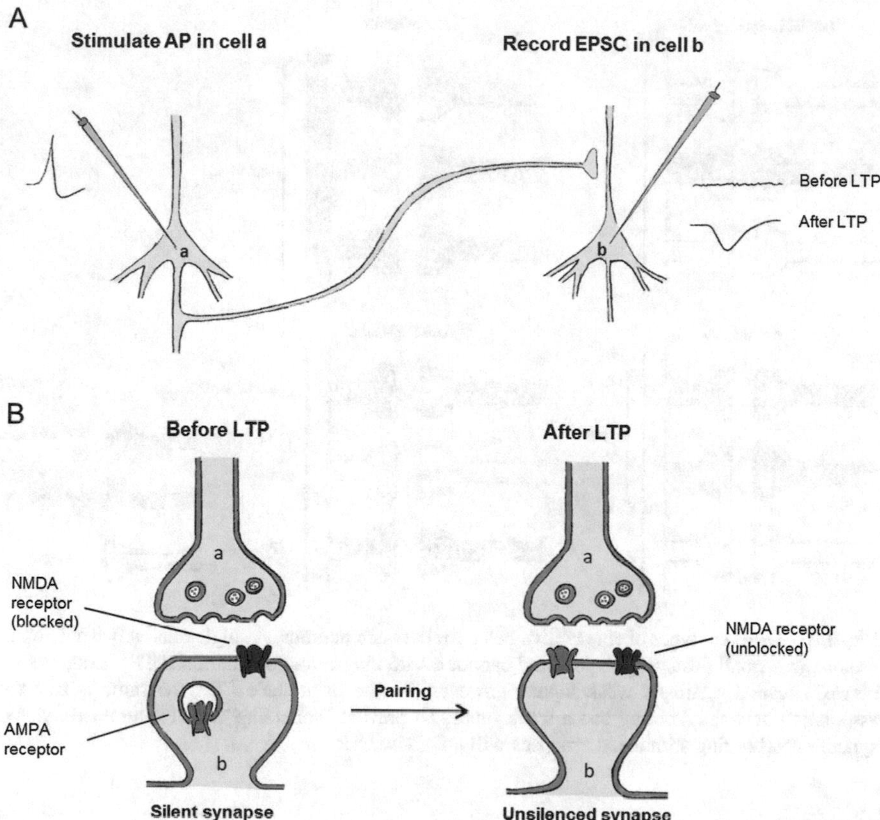

Fig. 1.26 Long-term potentiation can activate previously "silent", that is, nonactive synapses and in this way change processing in the network without formation of new synapses. (**A**) Stimulating neuron (a) connected to neuron (b) via a silent synapse does not produce an effect in neuron (b) before long-term potentiation (LTP), but does so after LTP. (**B**) The underlying mechanism is a deblocking of the NMDA receptor plus an activation of the AMPA-receptor

transmitters can be modified at synapses (we have omitted the quite rare electrical couplings between neurons) and just learned that "silent" synapses can be awakened. But synapses can also be produced "from scratch" and also be dismantled (see Fig. 1.29). The latter is most important during early childhood—it seems that the genes produce quite a few synapses that are of no use.

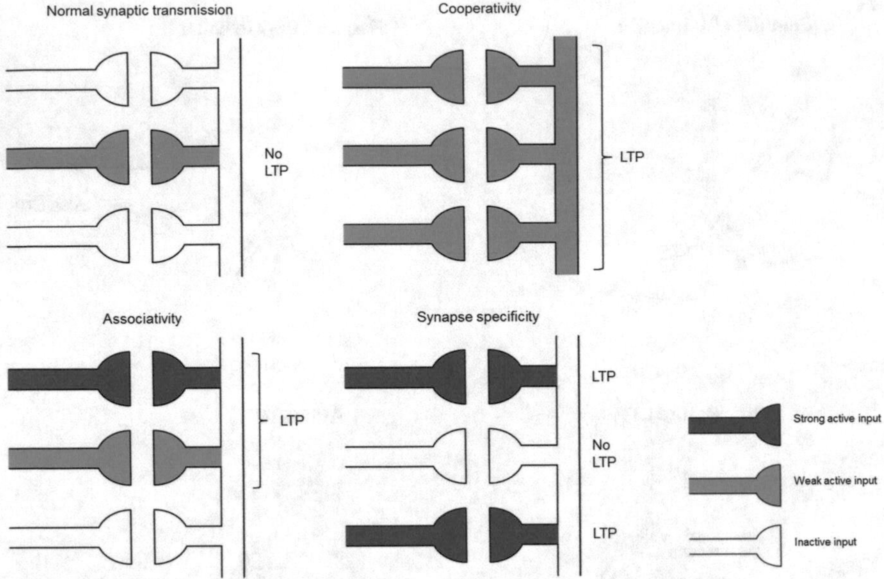

Fig. 1.27 Different types of cooperative behavior between neurons. A single relatively weak input, producing a small postsynaptic potential cannot evoke long-term potentiation (LTP). Cooperation between several relatively weak inputs, however, is able to produce LTP; the same is true for association between a strong and a weak input. To prevent "spreading" of LTP, an unstimulated synapse neighboring stimulated synapses will not show LTP

1.5 Modeling

A larger body of theoretical groundwork has been laid over the last decades regarding the storage of information in the brain. As indicated above, we now consider the brain as an assembly of interconnected neuronal networks that work in an associative way and are plastic, that is, can change as a result of prior experience. One early example comes from the work of Palm [21] on associative memories (see Fig. 1.30). While we do have detailed knowledge of the cellular and the subcellular mechanisms underlying information storage in nervous systems, the exact coding of information in these networks is not completely known. This is to say that we very well understand the workings of the brain on a cellular and on a subcellular level. We can also be sure that these well-understood elements are organized in neuronal networks, and we know relatively well which part of the cortical machinery does what (vision/hearing/somatosensing/motor). But we still do not know the consequences for network behavior of changes in the firing pattern of a single neuro or a small group of neurons. Work on "deep learning" in multilayered networks has led to great progress in artificial intelligence (AI) networks. There seems to be a close relationship between the functionality of these multilayered networks and the

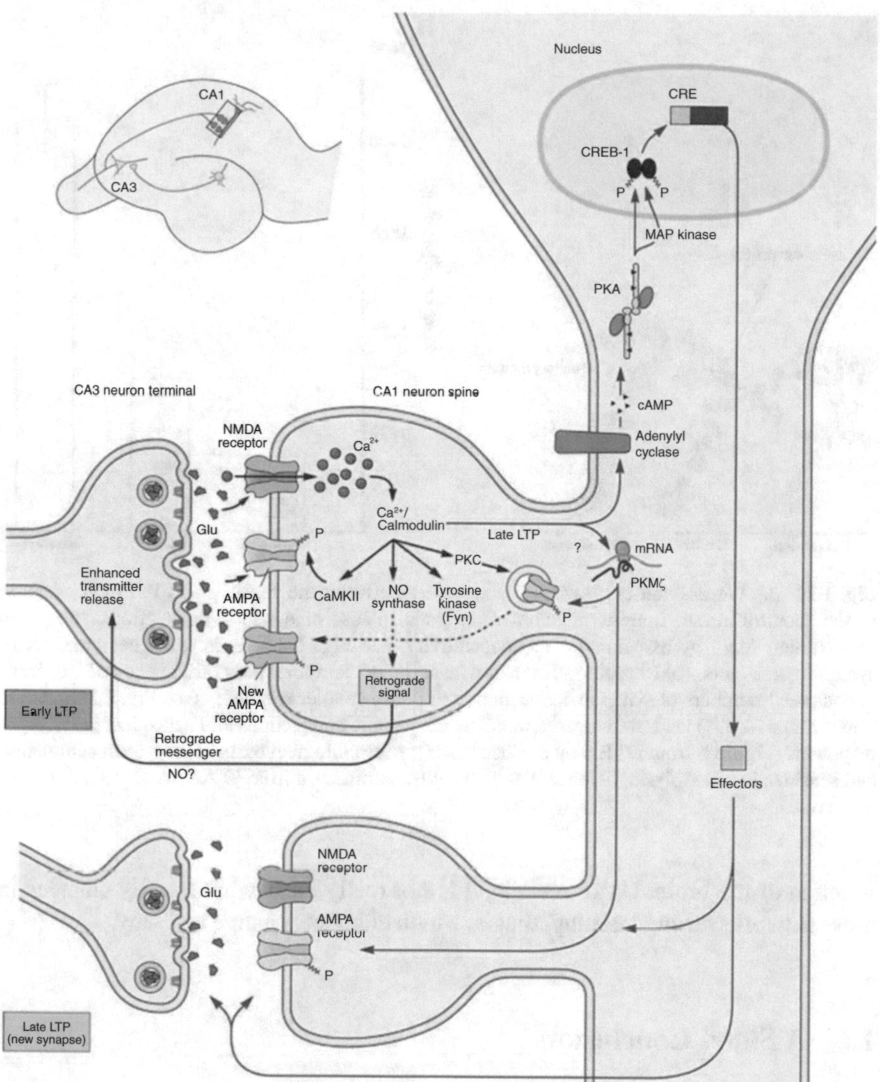

Fig. 1.28 A model for the molecular mechanisms of early and late phases of long-term potentiation. A single tetanus induces early LTP by activating NMDA receptors, triggering Ca^{++} influx into the postsynaptic cell and the activation of a set of second messengers. With repeated tetani, the Ca^{++} influx also recruits an adenylyl cyclase, which generates cAMP that activates PKA. This leads to the activation of MAP kinase, which translocates to the nucleus where it phosphorylates CREB-1. CREB-1 in turn activates transcription of targets (containing the CRE promoter) that are thought to lead to the growth of new synaptic connections. Repeated stimulation also activates translation in the dendrites of mRNA encoding PKMζ, a constitutively active isoform of PKC. This leads to a long-lasting increase in the number of AMPA receptors in the postsynaptic membrane. A retrograde signal, perhaps NO, is thought to diffuse from the postsynaptic cell to the presynaptic terminal to enhance transmitter release. (Figure and legend reproduced from [16] Kandel et al., 2013, p. 1502, with permission by © McGraw Hill Companies, 2018. All Rights Reserved)

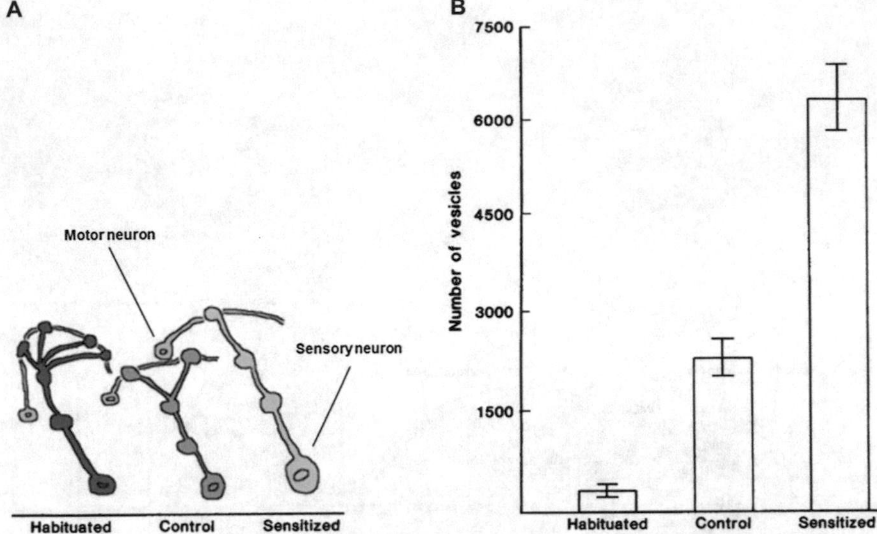

Fig. 1.29 (**a**) We assume two synaptic connectors between the sensory and the motor neuron in the "control" case. Long-term habituation leads to loss of synaptic connections, long-term sensitization (e.g., by associativity or cooperativity; see Fig. 1.27) leads to higher numbers of synaptic connectors. (**b**) Repeated stimulation (at sufficient temporal intervals, hence no LTP) leads to decreased numbers of synaptic connections and hence smaller responses (see Fig. 1.24b), while sensitization (see A) leads to increased numbers of synaptic connections and hence increased larger responses. (Figure **b** from [2] Bailey & Chen (1983): Morphological basis of long-term habituation and sensitization in Aplysia, Science 220, p. 92, with permission from © AAAS, 2018. All Rights Reserved)

function of the brain. Unfortunately, it is not really clear what exactly changes in these networks during learning, that is, which elements change and why.

1.6 A Short Conclusion

We know much about the function and limitations of our brain in terms of information intake, processing, and storage on both the behavioral and the (sub)cellular levels. But we still lack a detailed concept of how exactly these neuronal networks produce behavior, that is, how sensory input is analyzed, categorized, and evaluated to produce behavior. Hence, brain research is still a wide open field.

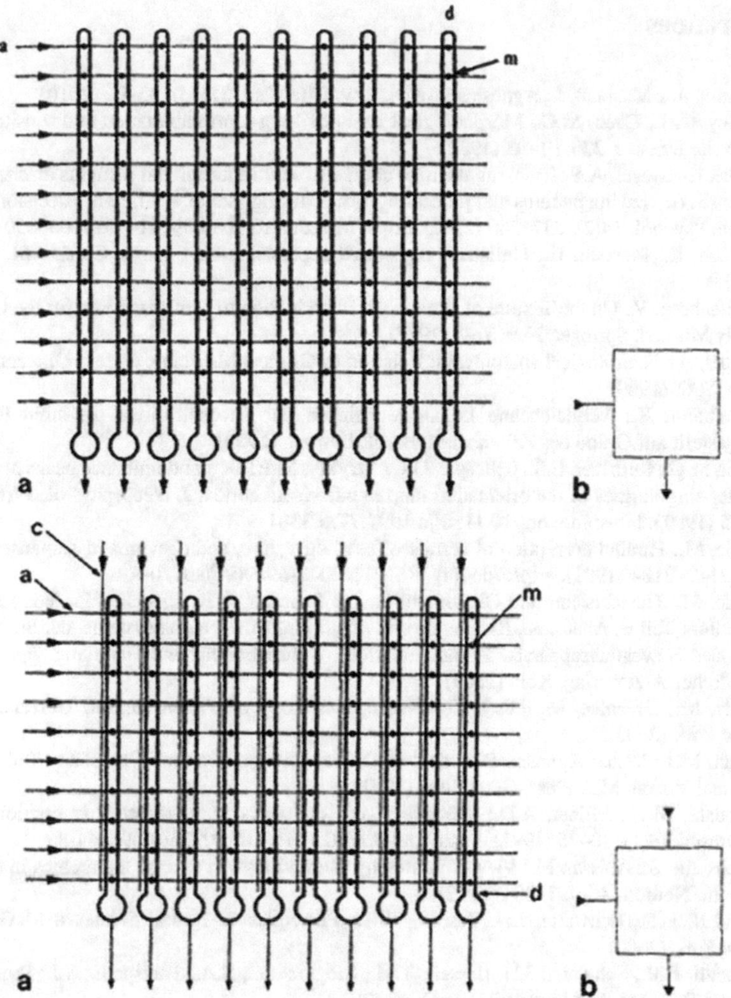

Fig. 1.30 Top: (**a**) A schematic neuronal realization. *a* axon; *d* dentrite; *m* modifiable synapse. (**b**) shorthand for a. Bottom: (**a**) A schematic neuronal realization of a recurrent associative neuronal network. *a, c* axons, *d* dentrite, *m* modifiable synapse. (**b**) shorthand for a. (Figures reproduced from [21] Palm (1982): Neural Assemblies. An Alternative Approach to Artificial Intelligence, p. 55, with permission from © Springer Nature, 2018. All Rights Reserved)

References

1. Álvarez, R., Masjuan, J.: Agnosias visuals. Rev. Clín. Esp. **216**(2), 85–91 (2016)
2. Bailey, C.H., Chen, M.C.: Morphological basis of long-term habituation and sensitization in Aplysia. Science. **220**, 91–93 (1983)
3. Belke, E., Meyer, A.S.: Tracking the time course of multidimensional stimulus discrimination: analyses of viewing patterns and processing times during "same"-"different" decisions. Eur. J. Cogn. Psychol. **14**(2), 237–266 (2002). https://doi.org/10.1080/09541440143000050
4. Bisiach, E., Luzzatti, C.: Unilateral neglect of representational space. Cortex. **14**, 129–133 (1978)
5. Braitenberg, V.: On the Texture of Brains. An Introduction to Neuroanatomy for the Cybernetically Minded. Springer, New York (1977)
6. Brodal, A.: Neurological anatomy. In: Relation to Clinical Medicine. Oxford University Press, New York (1969)
7. Brodmann, K.: Vergleichende Lokalisationslehre der Grosshirnrinde. In ihren Principien dargestellt auf Grund des Zellenbaues. Barth, Leipzig (1909)
8. Crair, M.C., Ruthazer, E.S., Gillespie, D.C., Stryker, M.P.: Ocular dominance peaks at pinwheel center singularities of the orientation map in cat visual cortex. J. Neurophysiol. **77**(6), 3381–3385 (1997). https://doi.org/10.1152/jn.1997.77.6.3381
9. Fahle, M.: Parallel perception of vernier offsets, curvature, and chevrons in humans. Vis. Res. **31**, 2149–2184 (1991). https://doi.org/10.1016/0042-6989(91)90170-A
10. Fahle, M.: Gesichtssinn und Okulomotorik. In: Schmidt, R.F., Unsicker, K. (eds.) Lehrbuch Vorklinik Teil B: Anatomie, Biochemie und Physiologie des Nervensystems, der Sinnesorgane und des Bewegungsapparats. Kommunikation, Wahrnehmung und Handlung, pp. 253–299. Deutscher Ärzteverlag, Köln (2003)
11. Fahle, M., Greenlee, M. (eds.): The Neuropsychology of Vision. Oxford University Press, New York (2003)
12. Farah, M.J.: Visual Agnosia. Disorders of Object Recognition and What They Tell Us about Normal Vision. MIT Press, Cambridge (1991)
13. Goodale, M.A., Milner, A.D.: Separate visual pathways for perception and action. Trends Neurosci. **15**(1), 20–25 (1992). https://doi.org/10.1016/0166-2236(92)90344-8
14. Hochstein, S., Ahissar, M.: View from the top hierarchies and reverse hierarchies in the visual system. Neuron. **36**, 791–804 (2002)
15. Kandell, E.R., Schwartz, J.H., Jessell, T.M.: Principles of Neural Sciences. McGraw-Hill, New York (2000)
16. Kandell, E.R., Schwartz, J.H., Jessell, T.M., Siegelbaum, S.A., Hudspeth, A.J.: Principles of Neural Sciences. McGraw-Hill, New York (2013)
17. Kanwisher, N.: Functional specificity in the human brain: a window into the functional architecture of the mind. PNAS. **107**(25), 11163–11170 (2010). https://doi.org/10.1073/pnas.1005062107
18. Kranz, F., Ishai, A.: Face perception is modulated by sexual preference. Curr. Biol. **16**(1), 63–68 (2006). https://doi.org/10.1016/j.cub.2005.10.070
19. Newsome, W.T., Britten, K.H., Movshon, J.A.: Neuronal correlates of a perceptual decision. Nature. **341**, 52–54 (1989). https://doi.org/10.1038/341052a0
20. Oxford Living Dictionaries. https://en.oxforddictionaries.com/definition/information. Accessed 4 July 2018
21. Palm, G.: Neural Assemblies. An Alternative Approach to Artificial Intelligence. Springer, Berlin (1982)
22. Penfield, W., Rasmussen, T.: The Cerebral Cortex of Man: A Clinical Study of Localization of Function. Macmillan, New York (1950)
23. Rey, A.: L'examen psychologique dans les cas d'encéphalopathie traumatique (Les problems.). Arch. Psychol. **28**, 215–285 (1941)

24. Roorda, A., Metha, A.B., Lennie, B., Williams, D.R.: Packing arrangement of the three cone classes in primate retina. Vis. Res. **41**, 1291–1306 (2001)
25. Salzman, C.D., Murasugi, C.M., Britten, K.H., Newsome, W.T.: Microstimulation in visual area MT: effects on direction discrimination performance. J. Neurosci. **12**(6), 2331–2355 (1992). https://doi.org/10.1523/JNEUROSCI.12-06-02331.1992
26. Treisman, A.M., Gelade, G.: A feature-integration theory of attention. Cogn. Psychol. **12**, 97–136 (1980)
27. van Essen, D.C., Anderson, C.H., Felleman, D.J.: Information processing in the primate visual system: an integrated perspective. Science. **255**, 419–423 (1992)

Chapter 2
Verbal Short-Term Memory: Insights in Human Information Storage

Michail D. Kozlov

Abstract It has often been suggested that verbal short-term memory, the ability to maintain verbal information for a brief period of time, is based on the upload of to-be-remembered material into passive, dedicated, information stores. Alternatively, it has been claimed that all information is remembered but that access to it gets obstructed because of interference by subsequent similar material. The aim of this chapter is to present both approaches and to examine the viability of a different, perceptual-gestural, view of information buffering over the short term. This approach conceptualizes verbal short-term storage as an active process that emerges from, and is defined by, the recruitment of receptive and (speech) productive mechanisms. Experimental results actually suggest an active involvement of productive mechanisms. These experiments also cast doubt on the proposal that forgetting occurs because of interference by similar content. Another experiment expands upon this challenge of the interference-based view by showing that a temporary lesion of a brain area involved in speech planning (Broca's area) affects verbal short-term memory performance in the absence of any additional potentially interfering verbal input. Further, challenging the store-based view, the virtual lesion of Broca's area also attenuated the phonological similarity effect, a hallmark effect of the function of the hypothetical language-independent store. Finally, based on further experiments, it is concluded that only the perceptual-gestural approach can offer an account of presentation-type-based differences in verbal list recall that goes beyond a redescription of the observed effects.

Parts of this chapter have been previously published in [55, 56], and in [61]. Methodological aspects of the research were included in [63].

M. D. Kozlov (✉)
Zentrum für Verhaltensmedizin, Luisenklinik, Bad Dürrheim, Germany

C. S. Große, R. Drechsler (eds.), *Information Storage*,
https://doi.org/10.1007/978-3-030-19262-4_2

2.1 Introduction

"HC SVNT DRACONES!" This Latin phrase, translated as "Here be dragons" was often used on ancient maps in order to denote unexplored regions. Naturally, if a region is unknown it is impossible to conclude that dragons do not live there. Whilst in medieval times this logic might have been sound, nowadays hardly anyone seriously believes that there are dragons living in the few unmapped regions of the globe. This assumption could be based on the inductive conclusion that since most of the earth has now been cartographed and dragons were not found, dragons won't exist in the remaining regions. However, black swans were also unheard of before Australia was discovered. Perhaps then it is premature to dismiss the existence of dragons. Yet, the nonbelief in dragons is founded on more than just the past inability to find them. There are also no fossil records, and it is difficult to conceive of a gland that would enable a living creature to spit fire out of its mouth. Dragons do not exist.

Just as dragons were proposed by medieval cartographers as explanations for why travelers rarely returned from their perilous journeys to foreign lands, many modern theories of verbal short-term memory, the ability to maintain verbal information in an active state over a brief period of time, often postulate bespoke mechanisms in order to explain short-term memory phenomena. This is partly because current knowledge about the human brain in many ways resembles medieval knowledge about the globe. Many brain functions have already been mapped but there is still much debate about the function of certain regions and the pathways connecting these regions are also yet poorly understood. Alas, adequate research techniques to map brain regions like functional magnetic resonance imaging or transcranial magnetic stimulation are still comparatively new and underdeveloped. It is therefore still possible to propose credible theories of mental entities like verbal short-term memory without giving much thought to whether it is plausible that those entities could be found in the brain. After all, perhaps the neural correlates of these entities have simply not been discovered yet. Surely, the label "Hic est animus," "Here be memory" should be appropriate somewhere in the brain.

As with dragons, in order to show that verbal short-term memory does not exist as a separate entity, two paths can be pursued: The first approach concerns mapping the brain. If it is impossible to find the short-term memory construct anywhere in the brain then it is narrow to assume that it does not exist. However, it could always be countered that the memory entity, like a black swan rather than like a dragon, is hiding in an undiscovered region. Until all regions and their connecting pathways are excluded as the seats of short-term memory this objection is difficult to refute.

The other approach is to demonstrate that the functions of already-mapped brain regions can sufficiently or better account for processes that are cited as evidence for the existence of a bespoke short-term memory system. By analogy, if it can be shown that other mundane dangers lie behind the horizon, it is not necessary to believe that foreign lands are dangerous because dragons live there. Thus, if existing perceptual and gestural processes that are clearly mapped onto brain regions can explain short-term memory phenomena, then parsimony dictates that it is not necessary to invoke bespoke short-term stores to explain the data. Note that in this case parsimony refers to evolutionary and not necessarily explanatory parsimony. In order to achieve explanatory parsimony for short-term memory phenomena it is easy enough to invent additional constructs such as an as-yet-unmapped memory store. This, however, is not much different from "parsimoniously" explaining that celestial bodies move because it is God's will. In contrast to this, evolutionary parsimony refers to how evolutionary pressures necessitate that any organ or appendage that does not serve a vital or reproductive purpose is selected against. This is because an appendage or brain region with no purpose would still require nutrients, which would have to be obtained by the organism through additional and potentially dangerous foraging. Incidentally, this is also the reason why the popular belief that humans only use 5% of their mental capacity is nonsensical [2]. Clearly, therefore, a brain region whose functions can be performed by other, more functionally generalized, mechanisms should have a very low likelihood of ever evolving. Hence, if it can be shown that basic perceptual and motor planning mechanisms can explain short-term memory phenomena, the assumption of additional memory storage entities becomes evolutionary unparsimonious.

The question of whether a dedicated verbal short-term memory system that can be mapped onto the brain exists is not trivial. All human communication is based on the assembly of meaningful utterances that form distinct and individually arbitrary gestures and symbols. Clearly, in order to write or speak or gesture a meaningful sentence in sign language, it is necessary that a long cascade of gestures is assembled into a coherent routine. Since the routine is issued over a temporally extended period it is necessary for some buffering to take place: Later words in a sentence have to be maintained in an active state while earlier words are produced. Thus, the process of short-term memory is vital for any meaningful verbal communication. Moreover, any goal-directed behavior that goes beyond stimulus–response associations requires the ability to meaningfully organize single actions into sequences [57]. Because the most sophisticated manifestation of such sequential behavior in humans is speech, an understanding of the formation and execution of speech sequences should be informative of the formation and execution of meaningful action sequences in general.

An important distinction needs to be drawn here between the *process* of short-term memory and the *mechanism(s)* of short-term memory. If short-term memory were accomplished by a dedicated system, then the process of short-term memory and the mechanism of short-term memory would be equivalent. Damage of this mechanism—as a result of a stroke for example—would necessarily impair the verbal short-term memory process. However, if the short-term buffering of verbal information is an emergent property of general receptive (e.g., perceptual) and productive (e.g., vocal-articulatory) mechanisms, then it would only make sense to speak of the short-term memory process but not of a short-term memory mechanism or system.

Another distinction between the conceptualization of verbal short-term memory as an emergent property of general language-related processes and the conceptu-alization of verbal short-term memory as the product of a dedicated short-term memory mechanism is the way in which the two approaches permit generalizing verbal short-term memory processes to other cognitive domains. If short-term buffering of verbal information is an emergent property of the interconnectivity of neurons in the verbal system, then it stands to reason that similar buffering can emerge from the interconnectivity of neurons inside the visual-spatial system, the kinesthetic system, or even inside a petri dish (c.f., [84]). Short-term memory performance in these domains would then resemble verbal short-term memory performance because in either case short-term memory would emerge from the same substrate. Indeed, some verbal short-term memory-like phenomena have been found outside the verbal domain (e.g., [59, 80]). If, however, the verbal short-term memory process were dependent upon the function of a verbal short-term store, then additional stores would have to be invoked for every domain in which short-term memory behavior is observed. This is implausible, because it is doubtful that the human brain could accommodate a bespoke store for every single activity that might need buffering. The alternative would be that a single or a limited number of stores accomplish all short-term memory functions in all cognitive domains. This assumption is also implausible. Arguably, a single or even a handful of stores would be overwhelmed by the buffering demands of the entire cognitive system. Nonetheless, if it should prove impossible to ascribe all verbal short-term memory phenomena to the function of general receptive and productive mechanisms, then the concept of a bespoke short-term store would have to be invoked in the verbal domain and in other domains, too.

2.2 Classical Theoretical Approaches to Verbal Short-Term Memory

2.2.1 Store-Based Approach

Perhaps the most prominent instantiation of the view that short-term memory function is supported by a bespoke system is the Working Memory model [4, 8]. The model postulates the existence of three storage systems: A phonological loop for short-term memory of verbal material, a visuospatial sketchpad for visual and spatial short-term memory, and an episodic buffer that acts as a short-term store for any information that is not stored by the other two systems. It should be noted that the episodic buffer is a relatively recent addition to the model [7]. Of the three buffer systems proposed by the Working Memory model, the phonological loop has received the greatest amount of research attention [8], and therefore lends itself particularly well for the discussion of the merits of a store-based approach to short-term memory.

The phonological loop is divided into two components, a passive phonological store and an active articulatory control process akin to subvocal speech [8]. As its name indicates, the phonological store holds to-be-remembered verbal items in a modality-neutral (and hence abstract) phonological code. Within the store, these phonologically coded items decay rapidly, and need to be refreshed via the articulatory control process (i.e., articulatory rehearsal), or else they are lost. Another critical feature of the phonological store is that despite the fact that its unit of currency is a modality-independent phonological representation, the route into the store differs according to the modality of input: If the to-be-remembered items are presented in the auditory modality, they gain direct, obligatory, access to the store. In contrast, visually presented material can only be uploaded into the phonological store indirectly via the articulatory control process. Thus, the articulatory control process has a dual function: To refresh decay-prone phonological representations (regardless of modality of input) and to convert graphemes into phonemes (for visual items).

It is a credit to the Working Memory model, that it identifies specific brain regions onto which the articulatory control process and the passive phonological store can be mapped, and thus avoids falling into the "dragon trap." Thus, it does not postulate that the dedicated short-term memory mechanism is out there somewhere, but makes specific and testable predictions about its location. The phonological store is suggested to be located in the left temporoparietal region of the brain, Brodmann Area (BA) 40, whereas the articulatory control process is mapped onto Broca's area (BA 44) [8]. Indeed, the Phonological Loop model even addresses the issue of evolutionary plausibility, arguing that the phonological loop evolved to facilitate language acquisition [3, 8, 10].

The Phonological Loop model is capable of explaining a broad range of verbal short-term memory phenomena. The standard test of verbal short-term memory is the verbal serial recall task. Participants are typically presented with a brief list of words, letters, or digits, which have to be reproduced immediately or very shortly after presentation, usually in their original order. This has long been the standard test of short-term memory because it is assumed to tap into the ability to organize sequences of actions, an ability that is central to much of goal-directed animal and human behavior, from locomotion, through reaching and grasping, to language use and the control of logical reasoning [57]. Variations of the to-be-remembered material, the modality of presentation and the addition of various concurrent tasks have revealed some very robust patterns of performance. These patterns are thought to reveal the limitation of human cognitive functioning, and any credible theory of short-term memory must therefore adequately account for these limitations.

Perhaps the most crucial benchmark phenomenon of verbal serial recall performance and one that is pivotal to the Phonological Loop model is the *phonological similarity effect*: It is more difficult to serially recall a list of similar "sounding" items, for example, "B," "C," "V," etc., than it is to serially recall a list of dissimilar "sounding" items, for example, "X," "Y," "Q," etc. [25]. The reason "sounding" is placed in quotation marks here is that, critically, this effect does not depend on whether the items are presented auditorily or visually [26], an observation that is central to the claim that the phonological store is indeed phonological: On the Phonological Loop model, the phonological similarity effect occurs because verbal items, regardless of input modality, gain access to an abstract-phonological store; the more similar these phonological representations, the more easily confused they are during retrieval from the store [5].

Another key phenomenon of verbal serial short-term memory is that the concurrent articulation of an irrelevant speech utterance like "the, the, the . . . " reduces serial recall performance markedly (also often termed "articulatory suppression"; [4, 13, 60, 67]). According to the Phonological Loop model, concurrent articulation blocks the loop's articulatory control process. Since this control process is needed to refresh decaying item representations in the store, concurrent articulation results in loss of items from the store and hence an overall performance reduction [14].

Crucial to the fractionation of the phonological loop into a passive phonological store and an articulatory control process is a three-way interaction between the phonological similarity effect, concurrent articulation, and input modality. If to-be-remembered material is presented visually then in addition to reducing overall performance, concurrent articulation reduces or abolishes the phonological similarity effect [12, 51, 66]. In contrast, with auditory presentation, concurrent articulation still negatively affects overall performance, but the phonological similarity effect is not fully abolished. Particularly in recency, that is, the last few items of a to-be-remembered list, the phonological similarity effect tends to be preserved under concurrent articulation [50, 51]. The Phonological Loop model accommodates this observation by pointing to the different pathways through which visual and auditory information gains access to the short-term store. Since auditory stimuli are obligatorily uploaded into the store, the phonological similarity of to-be-remembered items still determines the likelihood of inter-item confusion, even when the articulatory control process is suppressed. In contrast, if item presentation is visual, then suppression of the control process prevents items from accessing the store, so that their phonological similarity is immaterial for recall success.

A further canonical finding in short-term memory research is the effect of input modality. This *modality effect* manifests itself primarily as a recall advantage in recency on lists containing an auditory component when compared to pure visual lists. It is thus observable with pure auditory items (e.g., [51]), vocalized visually presented items [26], and visually presented items that are read to the participants [29]. Whilst the Phonological Loop model does not directly account for the modality effect, it is conceivable that the preferential access of auditory information to the phonological store proposed by the model somehow benefits recall on the auditory list.

Numerous findings from studies with brain damaged patients can also be accounted for by the Phonological Loop model. For example, in patients with a speech-planning impairment, like apraxia of speech, the phonological similarity effect is reduced for visually, but not auditorily, presented material [85]. In contrast, patients with peripheral motoric speech production impairments, like anarthria or dysarthria, show a normal phonological similarity effect [20]. The Phonological Loop model suggests that this is because the articulatory control process is based primarily on speech-planning mechanisms associated with BA 44. Therefore, a pathological disruption of the speech-planning mechanism disrupts the articulatory control process, which limits or prevents the access of visual information to the phonological store. As is the case under concurrent articulation, if items do not get access to the store then confusions between items on the basis of phonological similarity cannot take place, and the phonological similarity effect is reduced. Peripheral motoric speech impairment does not affect the control process and hence visual and auditory items still get access to the store where similar items are liable to be confused [8].

Finally, it has been observed that patients with damage to the left temporoparietal brain region tend to show selective verbal short-term memory deficits, in the absence of an immediately detectable impairment in speech fluency [81, 82]. This observation is central to the Phonological Loop model as it seems to clearly indicate that verbal short-term memory capacity is dissociable from general language-related processes. If verbal short-term memory can be selectively impaired through damage to a specific brain region, then this suggests both that the existence of a specialized language-independent verbal short-term memory mechanism is likely, and that this mechanism is located in the damaged region.

Alas, it is questionable whether the possibility of language impairments in "pure" short-term memory patients can be ruled out completely. It might always be the case that the language impairment is substantial enough to have a knock-on effect on verbal short-term storage, but not substantial enough to be detected by conventional tests of linguistic ability. Furthermore, neuroimaging studies with healthy volunteers have so far failed to confirm a specific region in the left temporoparietal brain area as the seat of the language-independent phonological store [22]. Hence, the existence of a neurological equivalent of a phonological store remains debatable.

The absence of unequivocal neurological evidence for a language-independent phonological store, however, begs the question whether it is necessary at all to invoke the theoretical concept of a short-term store as an entity in order to explain verbal short-term memory phenomena. One prominent alternative approach is to discard the idea that memory requires a dedicated store coupled with a separate active-refreshing process. Instead, it is suggested that all information is obligatorily remembered but that access to this memory is prone to interference.

2.2.2 Interference-Based Approach

Interference-based models of short-term memory are based on the assumption that there are two kinds of memories; a secondary memory in which all of a person's experience is stored, and a primary memory in which currently active representations are held [48]. Thus, no information is ever truly forgotten in secondary memory, but access to that information from primary memory is often occluded by interference. A prominent instantiation of interference-based short-term memory models is the Feature model [67, 68]. According to this model, each to-be-remembered item is composed of a number of modality-dependent features. Modality-dependent features are physical features of the item, like its visual shape or the voice in which it is spoken. When an item is encoded its features are simultaneously uploaded into primary and secondary memory. An additional set of internal modality-independent features is appended to the representation of an item in either memory. These modality-independent features arise from internal item-categorization processes. For example, if the same digit is presented twice, once auditorily and once visually, then the two memory representations of the digit will have many overlapping modality-independent features, but no overlapping modality-dependent features [67]. According to this model, forgetting occurs because items interfere with earlier items (retroactive interference) in primary memory. Specifically, a given feature of item n is overwritten if that feature is also present in item $n + 1$. Modality-dependent features interfere with modality-dependent features only, and the same applies to modality-independent features. Since retrieval depends on accessing the correct item from secondary memory given item features present in primary memory, the more degraded the representation of an item in primary memory, the more difficult it is to access the correct item in secondary memory.

The Feature model demonstrates how an interference-based model can successfully account for many verbal short-term memory effects without having to invoke the existence of a dedicated short-term buffer: For example, on this model, the phonological similarity effect occurs because similar items have more overlapping features, so that more retroactive interference between to-be-remembered items takes place in primary memory, making accurate retrieval of the correct items from secondary memory more difficult [67]. The modality effect is also easily explained by the Feature model, without any necessity for an additional acoustic store: Visually presented items are represented more heavily in terms of modality-independent features and auditorily presented items more heavily in terms of modality-dependent features. Since the modality-independent features are related to internally generated activity, it is more likely that the last visual item will be overwritten by internal activity like task-irrelevant thoughts or indeed by subvocal rehearsal of early list items [67]. This is because subvocal rehearsal of early list items reactivates the modality-independent features of these items so that they interfere with the modality-independent features of the last item. Hence, from the interference-based perspective, active rehearsal of list items is in fact considered somewhat detrimental for recall success. The Feature model explains the effect of concurrent articulation in a similar fashion by suggesting that the repetition of an irrelevant utterance activates modality-independent internal representations. These representations can interfere with modality-independent features of to-be-remembered material, which leads to a reduction in short-term memory performance. If the to-be-remembered material is highly confusable to begin with, such as when to-be-remembered items are phonologically similar, then the additional interfering features introduced by concurrent articulation will have less of an impact than when the to-be-remembered material consists of phonologically dissimilar items. This is how the Feature model accounts for the reduction of the phonological similarity effect in the presence of concurrent articulation. Since it is assumed that concurrent articulation generates primarily modality-independent-based interference and that auditory items are encoded to a greater extent in terms of modality-dependent features, suppression does not affect the phonological similarity effect as much if the to-be-remembered material is presented auditorily [68].

Thus, it seems that an interference-based account of short-term memory is also capable of explaining the effects of concurrent articulation, phonological similarity, and modality on verbal short-term memory without invoking the concept of bespoke short-term buffers. Nevertheless, a severe limitation of the interference-based approach to short-term memory is that very little concern is given to specifying what neurological equivalents there might be for entities like primary or secondary memory. For once, this makes the concepts of primary and secondary memory very "draconic." Clearly, these mental entities do not exist outside of the brain, that is, on a metaphysical plane, yet without any specification of their location inside the brain it is impossible to falsify the existence of these constructs. Believing in non-falsifiable entities is, alas, not much different than believing in dragons. In addition, the lack of specification of neurological correlates of the Feature model constructs reduces the utility of the interference-based approach for predicting the effects of neurological disorders, or indeed for explaining data associated with these. It is, for example, unclear from an interference-based perspective why damage to specific brain areas, like Broca's area, as observed in apraxia of speech [70], should affect verbal short-term memory performance in ways similar to concurrent articulation. It is difficult to see how damage to specific brain areas might, like concurrent articulation, introduce irrelevant item features that would interfere with items in primary memory. Indeed, if memory is conceived of as a passive process so that active maintenance of the to-be-remembered information through, for example, rehearsal is unnecessary or indeed disruptive, it is unclear how disruption of any brain mechanism might negatively affect memory performance.

2.2.3 *Perceptual-Gestural Approach*

The perceptual-gestural approach to verbal short-term memory [46, 47, 50, 51] is not as far removed from the store-based approach as interference-based models. Like the store-based account, the perceptual-gestural perspective argues that active maintenance of the to-be-remembered material needs to take place. A crucial distinction between the store-based and the perceptual-gestural perspectives, however, is that the former considers the maintenance process to be in service to a passive store whereas the latter rejects the idea of a bespoke short-term storage entity altogether. Instead, it is argued that verbal short-term memory is primarily an emergent property of the function of mechanisms that are not specifically mnemonic but ones involved in general perception and production processes. For example, the store-based tradition would explain the relationship between verbal short-term memory task performance and second language acquisition [3] with a bespoke short-term memory mechanism having evolved to facilitate language acquisition. In contrast, the perceptual-gestural account would suggest that the human ability to use language has evolved to facilitate human cohabitation, and that this ability can be recruited for short-term retention of verbal material. This is not to say, however, that the perceptual-gestural approach considers language as indispensable for verbal short-term memory, or short-term memory in general. Language is just exemplary of a very sophisticated ability that relies heavily on perceiving and gesture planning. Furthermore, it stands to reason that if information is categorized as verbal, it will be maintained in a verbal way. This is because any living organism needs to be economical with its energy expenditure [2]. Hence, if information is categorized as verbal then the linguistic neural path is likely to be the most well-trodden, and hence least effortful, for processing that information. For example, whereas an illiterate person might process a written letter as a visual token, a skilled reader is likely to encode the letter verbally.

In order to explain verbal short-term memory processes without invoking bespoke stores, the perceptual-gestural perspective emphasizes the planning of articulatory gestures and processes of (auditory) perceptual organization. In contrast to the Phonological Loop model, which only sees a role for articulatory mechanisms in refreshing the decaying representations of items in a passive phonological store, the perceptual-gestural perspective proposes that the articulatory plan itself serves as the repository of verbal information. Specifically, in order to maintain verbal information an articulatory motor plan is assembled wherein the to-be-remembered material is maintained as a series of articulatory gestures. The assembly and maintenance of the planned gesture sequence is however not flawless and transposition errors between to-be-remembered items are possible. These transpositions are akin to Spoonerisms and are more likely between items that require similar articulatory actions [50]. Additionally, the perceptual-gestural account draws attention to the high degree of sophistication and automation with which the perceptual system meaningfully organizes incoming pieces of information, particularly when the information is sequential. In the visual domain, the principles of perceptual organization have been described in the Gestalt literature (e.g., [54]). For example, it has been observed that the visual system tends to process continuous entities as cohesive objects [77]. These objects are, amongst other things, defined by their perceived edges, which constitutes the Gestalt principle of the figure–ground contrast. Similar principles apply in the auditory domain (see [21]), where certain characteristics of the auditory input like pauses or changes in voice are perceived as markers of distinct auditory perceptual objects or "streams" [35, 46, 47]. Evidently, any series of to-be-remembered events is subjected to a considerable amount of categorization and segregation, which is likely to influence its recall. Thus, perceptual organization influences the memory process before the to-be-remembered material could possibly reach any dedicated storage system. The perceptual-gestural account acknowledges this by proposing that the way the to-be-remembered information has been organized by the perceptual system influences the nature of the articulatory motor plan generated to maintain to-be-remembered information [46, 47].

According to the Phonological Loop model, verbal input is represented in a modality-neutral phonological code. In contrast, the emphasis on general acoustic and gestural processing of the perceptual-gestural approach suggests the codes are more peripheral and modality specific than "phonological." Thus, it is argued that the phonological similarity effect arises from a greater articulatory and not phonological confusability between items. This articulatory confusability leads to more frequent transpositions of the articulatory gestures through which the items are cohered into a sequence and maintained for serial recall [33, 50].

The perceptual-gestural account also redefines the impact of concurrent articulation. According to this account, suppression does not prevent the refreshing of decaying phonological item representations residing in some separate passive store but rather disrupts the formulation and maintenance of a gesture sequence assembled with the purpose of correct output of the to-be-remembered verbal material [50, 51]. This is because concurrent articulation itself requires the planning and production of verbal utterances, and thus limits the ability to recruit the articulatory planning system for the formulation and retention of a sequence-output plan. If the to-be-remembered material is not processed through the articulatory system, however, then the articulatory similarity of to-be-remembered items will have little impact on the likelihood of correct recall. Thus, the perceptual-gestural account explains why the phonological similarity effect is reduced under concurrent articulation.

Acknowledging the sophisticated perceptual streaming of an auditory list, the perceptual-gestural account is also capable of explaining the modality effect without invoking an additional store dedicated to retention of acoustic items. If in the auditory domain silences can serve as object-defining boundaries (cf. [21]), then it follows that the silence at the end of auditory to-be-remembered list presentation will act as such a boundary. The perceptual-gestural view suggests that the silence will thus act as a perceptual anchor, thereby facilitating the recall of the end of the auditory list [50]. It should be noted here that this principle does not apply with similar sounding items. Because similar items are less perceptually distinct, the transitions between the items are relatively indistinct, too. Thus, the perceptual boundary at the end of an auditory similar item list constitutes a less salient order cue, and auditory similar item list recall does not show a recency advantage. Moreover, because in the visual domain objects tend to be defined through spatial as opposed to temporal boundaries [21], the cessation of the presentation of the visual list is not as salient as it is for an auditory list and does not serve as such a strong anchor. Hence, performance in recency is superior for auditory lists.

Clearly, if the perceptual processes responsible for the modality effect are independent of processes responsible for the phonological similarity effect, that is, articulatory planning processes, then it is not surprising that these two effects should be observed independently of each other. Thus, for visual lists when the phonological similarity effect is attenuated with concurrent articulation recall of "phonologically" similar and dissimilar visual lists will be equal, because the end of the visual dissimilar lists does not constitute a perceptual anchor that improves recall. If to-be-remembered lists are presented auditorily, however, dissimilar lists will be recalled under suppression better than similar lists, but only because the auditory advantage in recency will not be affected by suppression. While this might seem like the phonological similarity effect is preserved under concurrent articulation, it is the perceptual-gestural view that the superior performance on auditory dissimilar lists in recency under suppression constitutes an acoustic not a phonological similarity effect [50].

Importantly, findings with brain-damaged patients can also be accommodated without postulating a dedicated short-term memory mechanism. Generally, since the perceptual-gestural account, like the Phonological Loop model and in contrast to interference-based models, argues that short-term retention of verbal information requires an active process it is conceivable that a lesion of the brain would impede that process and produce a short-term memory impairment. Thus, for example, the observation that patients with speech-planning impairments like apraxia of speech perform similarly to nonclinical experimental participants under concurrent articulation [85] is in line with the perceptual-gestural account. The account clearly predicts that if the speech-planning mechanism is impaired, then recall performance will be generally reduced, and the articulatory similarity between items should have no bearing on how well they can be recalled. This is because the account postulates that a speech plan needs to be assembled to maintain the to-be-remembered information. Thus, damage to the speech-planning mechanism is seen as direct impairment of the verbal short-term memory process rather than as an impediment of the process through which items are refreshed for storage in a separate bespoke system.

Finally, regarding the evidence for the existence of a brain area that could be regarded as the neurological equivalent of the phonological store [81, 82], it has been observed that the left temporoparietal region, the most probable location of the store, might instead be responsible for the integration of speech perception and production [44]. If verbal short-term memory emerged, primarily, from the function of speech perception and production processes, as postulated by the perceptual-gestural account, then it is clear that damage to an area responsible for the integration of these processes would result in a substantial verbal short-term memory deficit. The cause of this deficit would, however, not be the dysfunction of a bespoke storage mechanism, but rather the inability to upload the perceived verbal information stream into an articulatory motor plan for maintenance. The selectivity of the verbal short-term memory impairment (cf. [81, 82]) can also be thus accommodated: Selective impairment of a region integrating speech perception and production processes would not necessarily affect the discrete abilities to either perceive or produce speech. Only when integration of these abilities is required, as is the case when an articulatory motor plan is assembled from a perceived list of verbal tokens that needs maintaining, would a selective short-term memory deficit become apparent. Thus, while the left temporoparietal region might show the properties of a short-term buffer, it is clearly not an area that is language-independent and specifically dedicated to verbal short-term memory.

2.3 Empirical Approaches

It appears that store-based, interference-based and perceptual gestural approaches to explaining verbal short-term memory are about equally capable of accounting for the effects of phonological similarity, modality and concurrent articulation, and the interactions between them. From an evolutionary perspective, it seems that of the three presented accounts the perceptual-gestural is the most promising, because it postulates fewer dedicated system than the store based approach, and yet, in contrast to the interference-based view, enables a clear mapping of its constructs onto brain mechanisms. However, the principles of evolution are not yet entirely understood, and can often lead to surprising results (i.e., black swans). The likelihood of evolution of its constructs can therefore not be the sole criterion for dismissing a cognitive theory, in particular if the theory offers ways how its postulated constructs might be evolutionary plausible, like proposing that the phonological loop is a language-learning device [10].

Another approach is to empirically test the predictions of the short-term memory theories against new types of experimental manipulations. One particularly promising type of manipulation that will receive special attention in this chapter is the induction of brain lesions with transcranial magnetic stimulation. This technique makes it possible to temporarily reduce the activity of the stimulated region. Thus, it is possible to conduct lesion studies on healthy volunteers. Lesions studies can reveal a lot about the adequacy of short term memory theories. For example, the observation that patients with damage to speech-planning areas have severe impairments of verbal short-term memory (e.g., [85]) suggests that some active articulatory process is involved in maintaining verbal information. This is in line with the store-based and perceptual-gestural, but not with the interference-based view, as it is not clear how damage to articulatory planning should impair the function of primary or secondary memory. Lesions studies with patients are inherently problematic, however, because patients with appropriate lesions are rare, the brain damage is rarely selective, making it difficult to establish clear correlations between a single brain region and its function, and patients are often capable of compensating for their impairments. Inducing temporary lesions with transcranial magnetic stimulation circumvents many of these problems.

Following these deliberations, in the following, experimental results testing the predictions of the perceptual-gestural, the interference-based, and the store-based accounts are presented.

2.4 Insights from Empirical Studies

A clear distinction that can be drawn between the interference-based, store-based, and perceptual-gestural approaches to verbal short-term memory is the importance each approach attributes to articulatory-motoric planning processes. The perceptual-gestural account considers articulatory gesture planning processes as heavily involved in normal verbal short-term memory function. The Phonological Loop model also considers these processes as important, albeit regarding them as subservient to a passive short-term store. Both accounts, therefore, predict that even peripheral motoric impairment of articulatory processes should impede verbal short-term memory.

In contrast to this, from an interference-based perspective articulatory processes only play a role in the short-term maintenance of verbal information if they generate verbal representations. Thus, concurrent articulation impairs verbal short-term memory because the modality-independent features of the irrelevant verbal utterance interfere with the modality-independent features of the to-be-remembered material [68]. Nonverbal impairment of articulatory motor processes should hence have little effect on verbal short-term memory performance. In the next section, attempts to adjudicate between the contrasting predictions of the interference-based account, on the one hand, and the perceptual-gestural and store-based accounts on the other are described by investigating whether and how a nonverbal constraint on articulation, namely, chewing gum, impedes verbal short-term memory.

2.4.1 The Impact of Nonverbal Concurrent Tasks on Verbal Short-Term Memory

The store-based and perceptual-gestural accounts of verbal short-term memory suggest that any impairment (e.g., by concurrent oral activity) of articulatory planning/production processes will also impair verbal short-term memory. In contrast, the interference-based Feature model argues that a concurrent oral activity is only disruptive for short-term memory performance because it activates internal representations of verbal items that interfere with the to-be-remembered material. This section reports empirical evidence that chewing gum, a nonverbal constraint on articulation, impairs verbal short-term recall of both item order and item identity. Specifically, experiments showed that chewing gum reduces serial recall of letter lists. Chewing does not simply disrupt vocal-articulatory planning required for order retention: chewing equally impairs a matched task that required retention of list item identity. Manual tapping produces a similar pattern of impairment to that of chewing gum. These results pose a problem for verbal short-term memory theories asserting that forgetting is based on domain-specific interference.

Wilkinson et al. [86] found that when chewing gum, participants performed better on spatial item-recognition memory and short-term old/new number and word recognition tasks, compared to participants pretending to chew or participants without any instruction to chew. Additionally, when participants were only pretending to chew gum, their number recognition performance was still higher than that of the control group. However, on most other tasks—whether dependent on short-term memory or not—their performance was worse (for similar results, see [78]).

Beneficial effects of chewing gum have also been found for free recall of a relatively long list of words (15 items; [16, 49]). It has been suggested that the facilitative effects of chewing gum on memory may be mediated by an increase of blood flow to fronto-temporal brain regions due to the mastication process [86]. Others suggest that the effects might at least partly reflect a context effect, to which the flavor of the gum contributes rather than have to do with chewing or gum per se [16, 49].

At first glance, the lack of impairment as a result of chewing gum appears to be in line with the Feature model, and poses a problem for the store-based and perceptual-gestural accounts, which predict that any constraint on articulation should reduce verbal short-term memory performance. However, none of the previous studies that have examined the effects of chewing gum on short-term memory have employed serial recall, the bedrock on which theories of short-term memory have been built (e.g., [4, 8, 25]). The aim of the experimental series (for details, see [55]) presented in this chapter therefore was to dissociate between the conflicting predictions of the interference-based approach, on the one hand, and the store-based and perceptual-gestural approaches, on the other hand, through investigating the impact of nonverbal concurrent oral activity—chewing gum—on verbal short-term memory.

First, the effect of chewing (flavorless) gum on serial recall was tested experimentally (see [55], Experiment 1a). The result showed that chewing gum reduces verbal serial short-term memory performance. This finding is in line with the hypothesis that mouth/jaw movements that are not dedicated to articulatory planning should impair memory [8, 51]. From this standpoint, chewing movements may either disrupt encoding and refreshing of decay-prone phonological item representations (cf. [4]) or the assembly and maintenance of a motor sequence-plan (see, e.g., [46]). In any case, it seems that, against the predictions of the Feature Model, a non-verbal constraint on articulation can also impair verbal short-term memory.

The results of this experiment also indicate that the previous assertion that chewing gum is beneficial for short-term memory (e.g., [86]) must be qualified with an important caveat: In contrast to previous research in this area, when the task involves short-term memory for sequences of events as opposed to short-term item recognition or free recall, a clear reduction in performance is found as a result of gum chewing.

In sum, these results suggest that chewing gum, a nonverbal constraint on articulation, can reliably reduce verbal short-term memory performance. The effects of chewing thus resemble the effects of concurrent articulation. As it is difficult to see how chewing gum might generate modality-independent features, which might interfere with the modality-independent features of the to-be-remembered material, the present finding poses a challenge for the Feature model. However, if chewing gum were entirely like concurrent articulation, then it should also reduce the phonological similarity effect. This was not observed. If the action of chewing, however, differs from that of concurrent articulation, then it seems somewhat premature to dismiss the Feature model based on the present findings without investigating to what extent chewing and concurrent articulation are actually comparable.

One possibility is that the main effect of gum is simply not of sufficient strength to have the more subtle impact on the magnitude of the phonological similarity effect, with the main effect of concurrent articulation being typically much greater (cf., [4, 51, 66]). Several studies have observed that concurrent articulation has a particularly strong impact on *serial* short-term memory tasks, when compared to matched tasks not requiring memory for order (cf. [18, 60]). If the effects of gum were to match the effects of concurrent articulation, gum should also have a stronger impact on serial memory. The next section presents an experiment (see [55], Experiment 2) comparing the effect of chewing gum on a task requiring short-term memory for order with that on a matched task that requires the retention of item identity but not order.

A test of verbal short-term memory for a list of items that is devoid of the need to retain their serial order is the "missing item" task (e.g., [18, 23, 58]). Here, participants are required to identify a missing item from a randomly ordered fixed set of items (e.g., "7" is missing from the list "28149365" taken from the digit-set 1–9). Thus, each item presented must be retained so as to identify the item that is not. However, the serial order of the list items is immaterial and the task is not thought therefore to rely on sequence planning but rather on a judgment of item familiarity (e.g., [23]). A serial short-term memory task that is—other than the need to retain serial order—well matched to the missing item task is the probed order task [18, 47]. Here, participants are again presented with the randomized fixed set of items but at test are re-presented with one of the presented items (the probe) and required to indicate which item followed it in the list. This ensures that the missing item and the probed order tasks are matched on the stimuli and output requirements. If chewing gum is like concurrent articulation, then it should disrupt particularly tasks that require *serial* short-term memory. It should therefore adversely affect the probed order task more than the missing item task.

The results of this experiment ([55], Experiment 2) showed that, as in Experiment 1, chewing gum significantly impaired short-term memory for order as measured on this occasion by its disruption of probed order recall. However, this adverse effect extends to memory for item identity: Missing-item recall was compromised to a comparable degree to that of probed order recall. The adverse effects of chewing on short-term memory do not appear to be limited, therefore, to tasks that have typically been more strongly associated with articulatory sequencing [8, 46]. The results also show that chewing impairs short-term memory of visually and auditorily presented lists to a similar extent. This suggests that chewing is not impairing the kind of deliberate encoding often associated with visual as compared with the obligatory encoding of auditory lists (e.g., [4, 44, 46]). Instead, chewing seems to exert its effect at a more central stage potentially concerned with maintenance of the to-be-remembered material.

The results of this experiment provide mixed evidence regarding the implications of the effects of chewing gum for theories of verbal short-term memory. On the one hand, chewing, like concurrent articulation, consistently reduces short-term memory performance. Further, its effects are clearly not limited to a simple impediment of encoding; if this was so, it would not affect recall of visually and auditorily presented material to a similar extent. This is in line with short-term memory theories that invoke a key role for speech mechanisms and thus predict a negative impact of any task constraining articulation (e.g., [8, 51]). However, there is a discrepancy between the predictions of these theories and the present results insofar as they predict that impairment of speech-planning mechanisms that serve to maintain order information should impair the probed order task more than the missing item task. Moreover, the fact that concurrent articulation, by preventing rehearsal, reduces (indeed usually abolishes) the phonological similarity effect with visual presentation [13] but, as was noted earlier, chewing gum does not (Experiment 1), also militates against a simple account in terms of an impairment of speech mechanisms.

There are indications that the effects of chewing resemble more the effects of manual tapping than they do concurrent articulation. The tapping task traditionally involves the repeated placement of one or several fingers on a hard surface in a steady and rhythmic fashion. Chewing and tapping have both been suggested to promote cognitive abilities by releasing excessive muscle tension [36]. This assertion is challenged, however, by numerous studies demonstrating adverse effects of tapping on short-term memory (e.g., [41, 76]). Thus, a further experiment ([55], Experiment 3) replicates Experiment 2 in all respects except that chewing was substituted by simple tapping. The results showed no difference between the effects that tapping and chewing have on short-term item and order recall. The lack of a significant interaction between concurrent task and presentation modality further indicates that the adverse effects of neither tapping nor chewing are due to an impairment of item encoding. Rather, it seems that these peripheral motor tasks disrupt some modality unrelated process involved in the maintenance of items in a list regardless of whether the retention of their order is required. Because tasks that are thought to rely on vocal-articulatory sequencing to different extents (order-based tasks such as serial recall and probed order recall compared to the missing-item task; e.g., [18, 58]) are equally impaired by chewing and tapping, this maintenance process seems to be independent of such articulatory sequencing.

At first glance, the present data challenge short-term memory accounts that postulate that verbal short-term memory is a (negative) function of domain-specific interference. For example, the Feature model (e.g., [41, 67, 68]) suggests that concurrent irrelevant articulation reduces memory performance by generating task-irrelevant verbal representations that corrupt the representations of the (also verbal) to-be-remembered items. Clearly, if this were the case then a nonverbal constraint on articulation like chewing should have no impact on memory. However, proponents of the Feature model may appeal to a free parameter included in the model representing a general attentional resource (parameter “a”). Thus, nonverbal concurrent tasks can impair short-term memory because they increase general task demands and deplete attentional resources needed for successful item retrieval [41, 68]. Tapping and chewing may therefore both simply be general distracters. Indeed, both concurrent tasks had very similar effects on verbal short-term memory, and neither of the tasks produced results that would usually be associated with a constraint on articulation, like reduction of the phonological similarity effect or selective impact on memory for order. It seems therefore on second glance that the Feature model offers in fact the best explanation for the current data.

That said, the "general distraction" explanation seems questionable. First, the Feature model does not seem to offer a way to determine a priori the extent to which a task should deplete the general attentional resource. Indeed, invoking the Feature model, it would be impossible to predict that the simple tapping task should produce the same amount of distraction as the, physically very different, chewing task (cf. [52]). Moreover, by invoking the parameter "a," the model implies that concurrent tasks that convey phonological features, like concurrent articulation, and concurrent tasks without a phonological component, like tapping or chewing, impact verbal short-term memory through different mechanisms. Inspecting the present results in conjunction with related literature, it then becomes unclear why tasks with and without a phonological component produce such similar results. For example, complex tapping, like concurrent articulation, can reduce the phonological similarity effect (see [41]), and steady-state concurrent articulation, like simple tapping or chewing, does not have a distinctive impact on serial memory tasks [60]. These similarities between verbal and nonverbal impairments of short-term memory suggest that the degree of impairment is determined by the complexity of the planned gestures involved in the concurrent task, not, as the Feature model would predict, its similarity to the to-be-retained material.

With the Feature model not offering a satisfying explanation for the present data, another glance at verbal short-term memory accounts that invoke a key role for language planning/production processes [1, 8, 51] is in order. The majority of these accounts differentiate between constraints on articulatory planning and those on articulatory production. Indeed it has been demonstrated that patients with anarthria, an impairment of the neuromuscular mechanisms required for articulation, show no reduction of the phonological similarity effect [15]. Only when patients show speech-planning deficits, as opposed to pure production deficits (such as in apraxia of speech), is a clear reduction of the phonological similarity effect observed [85]. Similarly, steady-state suppression, that is, concurrent articulation with low speech-planning demands, like the concurrent repetition of a single letter, reduces performance on the missing item task and the probed order task to a comparable extent [60]. Only changing-state suppression—concurrent repetition of a sequence of, say, three letters—reduces performance on the serial memory task more than on the missing item task. Thus, accounts that see a central role for language planning/production processes [8, 51] can be reconciled with the present findings if it is assumed that both chewing and tapping impair articulation at a peripheral level. At that level, the concurrent activity reduces overall performance but does not differentially affect performance on phonologically similar and dissimilar lists, nor differentially affect performance on order and item recall tasks. Thus, from this standpoint, tapping and chewing are not simply distracters: They are peripheral impairments placed on the production aspect of the articulatory planning and production network needed to either refresh decaying item representations in a short-term store (e.g., [8]) or to assemble a coherent motor-plan for action (e.g., [46, 47, 50]).

2.4.2 *Theta-Burst Stimulation of Broca's Area Modulates Verbal Short-Term Memory*

In this section, the interference-based, store-based, and perceptual-gestural approaches to verbal short-term memory are evaluated, by empirically addressing the varying predictions the accounts make about the consequences of speech-planning impairment. Because the Phonological Loop model identifies BA 44, Broca's area, [8] as the location of the articulatory control process, the model predicts that a lesion to the area should reduce visual verbal short-term memory performance and attenuate the phonological similarity effect. Without the articulatory control process, visual material should not gain access to the phonological store. Thus, visual items would not be maintained irrespective of their phonological similarity. Overall, reduced function of Broca's area should have effects similar to concurrent articulation.

At first glance, the perceptual-gestural account makes similar predictions about the consequences of damage to Broca's area for verbal short-term memory. The account suggests that inhibition of the articulatory planning mechanism should impair the ability to assemble an articulatory plan for visual to-be-remembered list recall. Consequently, recall performance should be reduced as should the likelihood of articulatory confusions and hence the phonological similarity effect. Yet, it should be noted that the perceptual-gestural account does not explicitly specify Broca's area as the seat of the speech-plan assembly mechanism. Indeed, given how the account usually emphasizes the interaction of perceptual and speech-planning processes, which involve a large number of brain areas, it seems more in line with the account to consider Broca's area merely a component of a distributed mechanism capable of generating an articulatory plan. This means that, according to the perceptual-gestural view, a selective lesion of Broca's area might produce a very selective impairment of the speech-plan assembly process, reducing, for example, only the likelihood of articulatory confusions or only the likelihood of correct recall. Finally, from an interference-based perspective, a selective lesion of Broca's area should have no effect on short-term memory performance because, in contrast to concurrent articulation, a lesion would not introduce interfering item representations to the memory traces of to-be-remembered items. In this section, these varying predictions are addressed empirically by applying repetitive transcranial magnetic stimulation to the pars opercularis of the left inferior frontal gyrus in order to induce a temporary lesion of Broca's area in healthy volunteers ([55], Experiment 4). Activity of a brain region associated with the articulatory control process, Broca's area (localized through a combination of structural and functional methods), was inhibited with theta-burst transcranial magnetic stimulation. According to the Phonological Loop model, this temporary lesion should reduce access of visual-verbal material to the store, resulting in a deficit in overall short-term memory performance. However, this was not observed; rather, there was a selective attenuation of the phonological similarity effect. This dissociation of the effect of transcranial magnetic stimulation of Broca's area on overall performance and on the impact of phonological similarity seems more readily accommodated by accounts that emphasize a primary role for articulatory processes in serial short-term memory rather than ones that regard such processes as peripheral to a dedicated store.

To recap, the most prominent store-based short-term memory account, the Working Memory model [8, 11], posits that in the verbal domain short-term retention is accomplished through the action of a phonological loop, which comprises two components: a bespoke, passive, language-independent *phonological store* in which phonological representations of verbal input last for one or two seconds before decaying, and an *articulatory control process*, a rehearsal mechanism analogous to subvocal speech, which serves to reactivate the stored items, thus preventing their decay. The articulatory control process is also the means by which visually presented verbal material gains access to the phonological store. Auditory items, on the other hand, have direct access to the store [8].

These key propositions of the Phonological Loop model account for a wide range of empirical phenomena, chief among them the phonological similarity effect [4, 25]. According to the Phonological Loop model, phonologically similar items are more readily confused inside the store [5], leading to poorer recall. Another canonical effect explained by the Phonological Loop model is the impact of concurrent articulation: The model suggests that the concurrent production of an irrelevant verbal sequence disrupts the articulatory control process, which means that the phonological representations of the to-be-remembered items cannot be refreshed, and are more readily lost. Moreover, for visual to-be-remembered material, disruption of the articulatory control process impairs access to the phonological store. Thus, the most empirically obvious impact of concurrent articulation with visual lists is an impairment of overall recall performance (e.g., [14, 66]). A secondary consequence of the impairment of access of visual material to the phonological store is the reduction of the phonological similarity effect with visual lists during concurrent articulation (e.g. [7, 9, 12, 13]): When items cannot enter a store designed specifically to hold verbal items, verbal recall is impaired generally (otherwise it is unclear why such a store would have evolved at all), but this impairment is especially pronounced for items that would, through being phonologically discriminable, have particularly benefitted from gaining access to the store.

In recent decades, neuroimaging and neuropsychological evidence has been brought to bear on the Phonological Loop model. Studies with speech-impaired patients have shown that peripheral motoric impairments of speech production, as observed in anarthric and dysarthric patients, do not impact upon effects like the phonological similarity effect [20]. However, apraxic patients—those with a deficit in speech *planning*—lack a phonological similarity effect for visual but not auditory lists [85], exhibiting a pattern of performance similar to that of nonclinical participants under concurrent articulation. Thus, the articulatory control process within the Phonological Loop model has been pinpointed to Broca's area (BA 44), the area that is commonly damaged in apraxic patients [70]. Working Memory model-inspired imaging studies found this area to be active during tasks that supposedly tap into the function of the articulatory control process of the phonological loop [71]. Furthermore, the passive phonological store component of the model has been mapped onto BA 40 [8] based on brain-damaged patients in whom damage to this area seems to have resulted in a selective, "pure" impairment in verbal short-term memory tasks in the absence of a substantial general language impairment (see [81] for a review).

In combination, the cognitive and neurological aspects of the Phonological Loop model allow clear predictions about the function of several brain areas and the consequences of lesions to these areas. Thus, the model predicts that lesions to BA 40, the phonological store, will result in a reduction in verbal short-term memory performance. A further consequence of such impairment is a reduction of the phonological similarity effect: Without a mechanism to store phonological representations of to-be-remembered items, the phonological similarity of the items ceases to be relevant for recall success. Selective lesion of BA 44, that is, damage to the articulatory control process, should have similar results but only for visually presented items, because the control process is the pathway through which these items gain access to the phonological store. Given that it should be immaterial whether access to the store is blocked because the store itself is damaged (lesion of BA 40) or because access to an (intact) store is constrained (lesion of BA 44), either form of selective impairment should lead to a reduction in short-term memory performance and a reduction of the phonological similarity effect, at least for visual to-be-remembered material.

From the store-based perspective, the objection could be raised that the difficulties with finding clear neurological correlates of the phonological loop might be inherent to the interpretational difficulties associated with single-case neuropsychological data and correlational imaging data. One way in which these difficulties might be circumvented, however, is to use transcranial magnetic stimulation. In particular, given that this technique can be used to temporarily induce lesions in the brains of healthy volunteers it is possible to test a sample drawn from a known population that is homogenous, thereby avoiding the risk of sampling error as might be brought about by a range of other factors, such as medication, socioeconomic class, age, and so on. A further major advantage of transcranial magnetic stimulation is that participants can serve as their own controls [75]. Moreover, the potentially confounding effects of auditory and tactile artifacts that accompany transcranial magnetic stimulation can be avoided by using an offline protocol in which transcranial magnetic stimulation is applied prior to the performance of a behavioral task. In the study reported here ([55], Experiment 4), therefore, the technique of continuous theta-burst stimulation was used [45]. This high-frequency, low-intensity protocol produces a suppressive aftereffect on cortical excitability for up to 1 h and beyond (e.g., [83]).

Like the Phonological Loop model, the perceptual-gestural account [46, 47, 50, 51], in regarding verbal short-term memory an emergent property of perception and speech planning, considers Broca's area to be involved in the short-term memory process, given that the area is commonly associated with speech planning (e.g., [31]). Yet the account does not specify Broca's area as the sole seat of the speech-plan assembly process. It is rather in line with the account to consider Broca's area as one important node in a network of brain areas responsible for speech production, damage to which should impact upon but not necessarily disrupt the verbal short-term memory process.

In order to address the predictions of the models, Broca's area was carefully localized in a sample of volunteers ([55], Experiment 4), first structurally on their MRI scan then functionally, by applying repetitive transcranial magnetic stimulation to various locations within the identified region until a speech-planning arrest hotspot was found. There, a theta burst of transcranial magnetic stimulation was administered, inhibiting the activity of the brain area for about 30 min. Afterward, participants were given a visual-verbal short-term memory task. Contrary to the predictions of both the interference-based accounts and the store-based account, theta-burst stimulation of Broca's area reduced the phonological similarity effect while average performance remained unaffected. Thus, it is not clear from an interference-based perspective how a virtual lesion of a brain area associated with speech should have either introduced interfering verbal features or depleted a central attentional resource. Moreover, from the store-based perspective of the Phonological Loop model, the primary effect of inhibiting Broca's area should have been a reduction in average performance and not an isolated reduction of the phonological similarity effect. Only the perceptual-gestural approach can accommodate the selective effect that temporarily lesioning Broca's area had on verbal short-term memory. This is because the approach, while suggesting that Broca's area is important for the verbal short-term memory process due to its involvement in speech planning, does not postulate that the area is the seat of the process. This allows for the occurrence of peculiar behavioral effects like a phonological similarity effect reduction in the absence of a reduction of overall performance arising from a stimulation of Broca's area.

Whereas theta-burst stimulation reduced the phonological similarity effect, it did not reduce overall performance. While performance on dissimilar lists decreased, performance on similar lists increased. This finding echoes previous research demonstrating that transcranial magnetic stimulation of a region assumed to be the seat of the phonological store reduced the phonological similarity effect by improving performance on phonologically similar items [53]. However, if, as the Phonological Loop model postulates, the articulatory process acts as a gateway through which visual-verbal material gains access to the store, then inhibiting the articulatory process with theta-burst stimulation should restrict the gateway to the store and thus reduce the ability to upload to-be-remembered material into it. With access to the mechanism that is indispensable for short-term maintenance of verbal material restricted, the primary consequence should be a reduction of short-term memory performance. Reduction of the phonological similarity effect would then arise as a secondary consequence, because a restriction of access to the phonological store would limit any advantage to-be-remembered items might gain though being phonologically discriminable. Clearly, this store-based account struggles with the present observation that the reduction of the phonological similarity effect and the reduction in verbal short-term memory performance can be dissociated.

While the current results present a challenge for the Phonological Loop model, it is possible that they may be accommodated within the broader framework of the Working Memory model. For example, it might be argued that theta-burst stimulation of the control process caused participants to abandon the use of the phonological loop and rely on other mechanisms. Within the confines of the Working Memory model they could have thus recruited the visuospatial sketchpad, a short-term buffer for visuospatial information, like graphemic representations of the to-be-remembered items [7, 11]. Participants could have also recruited the episodic buffer, a universal storage device invoked to explain short-term storage that cannot plausibly be accomplished by the phonological store or the sketchpad [8]. Thus, theta-burst stimulation could have induced a tendency to abandon phonological processing of the to-be-remembered lists. If the to-be-remembered lists are not encoded phonologically, then it follows that the impact of phonological similarity on recall would be reduced, but, if the alternative mechanisms are equally efficient for serial recall, overall performance might remain intact.

The problem with the phonological store-abandonment idea, however, is that it raises the question of why there should be a bespoke mechanism for phonological short-term storage if relying on other multipurpose mechanisms is equally efficient. Furthermore, one has to wonder whether in patients with a speech-planning impairment, their lesion also generally affects the ability to recruit alternative mechanisms for short-term memory, and if it does not, why these patients are unable to compensate for their impairment of the phonological loop, like the participants in the study of Kozlov ([55], Experiment 4). In sum, these results present substantial challenges to the Working Memory model, and in particular its phonological loop construct. It is for this reason that we look to alternative accounts of short-term memory to explore whether they offer a better fit for the present data.

Following the arguments put forward above, it is clear that the interference-based approach to verbal short-term memory, at least as instantiated in the prominent Feature model [19, 67, 68], is also incapable of accounting for the present data. The key point for present purposes is that the Feature model assumes that concurrent articulation interferes with performance because additional irrelevant item representations are introduced into primary memory and interfere with the representations of the to-be-remembered items [68]. The articulatory action itself is argued to have little bearing on the memory trace. Kozlov ([55], Experiment 4), however, clearly suggests otherwise: short-term memory performance was modulated with theta-burst stimulation of a speech-planning area in the absence of any additional item features being introduced to the memory trace. Note that the argument that such constraints might deplete an attentional resource, also does not account for the present results: It is difficult to see how theta-burst stimulation administered at the beginning of an experimental session to the speech-planning area should deplete attention. Moreover, depletion of an attentional resource could not account for the observation that performance on dissimilar items decreased, while performance on similar items increased. It seems, therefore, that interference-based models cannot account for the data any better than the Working Memory model.

Finally, according to the perceptual-gestural account [46, 47, 50, 51], it is not necessary to invoke bespoke short-term buffers to account for serial short-term memory phenomena. For example, it has been demonstrated that concurrent articulation reduces the phonological similarity effect for auditory lists just as much as for visual lists, except for the last few items in the auditory list [50, 51]. This recency advantage, however, is not due to obligatory phonological storage of the auditory list, but is based on sensory-acoustic factors governing the sequential perceptual organization of the auditory list that are not in play in the case of visual lists (e.g., [51]). That the key signature of the phonological store—the phonological similarity effect—is absent regardless of modality when rehearsal is impeded by concurrent articulation obviates the need to posit an additional passive store to which auditory information has preferential access. Instead, the phenomena of verbal short-term memory such as the phonological similarity effect are, primarily, products of the articulatory planning process itself. In this view, the phonological similarity effect results from exchanges between articulatorily similar elements, akin to Spoonerisms, during the speech-planning process ([50], see also [1, 33]). Indeed, without a default assumption of phonological storage, the present finding that the phonological similarity effect was reduced as a result of inhibiting articulatory planning would suggest a clear link between the similarity effect and articulatory processes.

By not invoking a dedicated storage mechanism, the perceptual-gestural account need not be committed to the idea that verbal short-term memory is associated with any single brain area. Instead the short-term memory process is likely to be distributed across areas involved in perception, action planning and production, and the integration of perception and action. Given the nature of the material, for verbal short-term memory the areas that are recruited are generally language-related. This includes BA 44, and BA 40, areas that are considered to be the locations of the articulatory control process and the phonological store, respectively, by the Phonological Loop model. Thus, it is in line with the perceptual-gestural view, as much as it is in line with a store-based view, that damage to BA 44 or BA 40 should impair verbal short-term memory. Since BA 44 is associated with speech planning and BA 40 is associated with the integration of perception into speech-related action [44], both are important areas for the assembly of a motor-plan for reproducing to-be-remembered verbal material. Yet, to assume that selectively impairing either area should be sufficient to disrupt the entire verbal short-term memory process seems too restrictive and localized to be in line with the picture of verbal short-term memory as an emergent property of receptive and productive mechanisms drawn by the perceptual-gestural account (e.g., [50]). If Broca's area is considered merely one of many areas contributing to the verbal short-term memory process and the verbal short-term memory process is not predicated on the function of the area (e.g., because it is the primary pathway for visual-verbal material into a bespoke storage mechanism) then it is conceivable that selective inhibition of the region could have a very selective effect on the short-term memory process. Thus, selective impairment of Broca's area could plausibly affect the verbal short-term memory process in a way that would simultaneously improve performance on similar items and reduce performance on dissimilar items. It might be possible, therefore, to account for the results of the presented study ([55], Experiment 4) from the perspective of the perceptual-gestural view, although further research will be needed to identify the details of such an account.

It is important to emphasize that the perceptual-gestural account suggests that the recruitment of articulatory mechanisms for short-term recall is task-driven and opportunistic. Thus, it is likely that some participants could opt for less articulation-dependent strategies to maintain the to-be-remembered list. While these participants would show a reduced phonological similarity effect, it does not need to be associated with an overall reduction in performance, the pattern observed in the present study. This is because these participants would, according to the perceptual-gestural account, simply choose a different strategy for list maintenance and not, as the Phonological Loop model would claim, abandon a bespoke mechanism required for short-term storage.

2.4.3 A New Approach to Modality Effects in Verbal Serial Recall: Meeting the Challenge of Explaining a Visual Mid-List Advantage

One major focus of this chapter is comparing and evaluating the predictions of the three prominent verbal short-term memory accounts in regard to the effect of perceptual factors on verbal short-term recall. In the past, the center of such a discussion would be the standard modality effect, the auditory advantage in recency when comparing visual and auditory list recall. Indeed the standard modality effect can be accommodated by any of the three prominent verbal short-term memory accounts. Thus, the interference-based account postulates that auditory items are encoded more in terms of modality-dependent features that are not prone to interference from internal activity at list end. The store-based view explains superior performance in recency on auditory lists with preferential access of auditory information to the phonological store, or invokes an additional low capacity buffer dedicated to storing auditory items. The perceptual-gestural view argues that the silence at the end of the auditory list constitutes a perceptual anchor that improves recall in recency.

In this section, the focus shifts toward comparing the predictions that various verbal short-term memory accounts make about the impact of presentation modality on memory for item sequences. Many verbal short-term memory theories propose an inherent advantage for recall of auditorily presented sequences. For example, the store-based Working Memory model [6] suggests that auditory-verbal items have direct access to a bespoke phonological short-term store. The store-based perspective has also proposed the existence of bespoke *acoustic* stores, like the Precategorical Acoustic Store [30], a limited capacity buffer dedicated exclusively to the retention of auditory information. It has also been claimed that a sequence of auditory items is encoded with greater positional resolution [43]. The interference-based Feature model [67, 68] also assumes an inherent advantage for auditorily presented material. According to the Feature model, to-be-remembered items are represented in memory in terms of a mixture of modality-dependent physical features and modality-independent features arising from internal processing of the items. The model argues that auditorily presented items are encoded primarily in terms of modality-dependent features. Thus, representations of auditory items in primary memory are less prone to interference from internal processes. The perceptual-gestural account [46, 47, 50, 51], points out—with reference to findings on the perceptual organization of sound into auditory streams [21]—that an auditory-verbal to-be-remembered item sequence tends to be perceived as a temporally extended object, with the silence at the end of the sequence demarcating the object boundary. Additional accounts of verbal short-term memory that focused primarily on explaining the modality differences have argued that auditory to-be-remembered material is encoded in an acoustic code so that its maintenance requires less allocated attention than the maintenance of visual material [72, 73]. There have also been claims that auditory items are encoded with better temporal resolution than visual items [37].

All these theories are confirmed by the robust and frequent observation (see [73], for a review) that, particularly at the end of a to-be-remembered list (i.e., at "recency"), items with an auditory component are remembered better than visual items: the *modality effect*. This effect can be observed if participants are required to read visually presented items out loud [26], if visually presented items are read to the participant [29], or if items are purely auditory (e.g., [51]). At first glance it seems therefore, that there is indeed, as many verbal short-term memory accounts claim, a hardwired benefit to memory if presentation of to-be-remembered material is auditory.

There are, however, several stumbling blocks for theories claiming an inherent memory advantage for auditory presentation. One is that a recency advantage is also often obtained with to-be-remembered lists that do not contain an acoustic component, like lists of visually presented verbal items that are silently mouthed (i.e., gestured without being vocalized), or lip-read lists [40]. Such findings raise the question of whether the recency advantage might be associated with the way to-be-remembered sequences are processed, as opposed to the modality in which they are presented per se.

Another obstacle for the assumption of a hardwired auditory advantage is the inverted modality effect: The auditory advantage in recency is often matched by a visual advantage in pre-recency, the early to middle portion of the serial position curve. This effect has been often overlooked, having been the object of research only rarely [17]. If the effect is genuine, and the auditory recency advantage is indeed often matched by a visual pre-recency advantage, then this calls into question any claim for a dedicated cognitive or neurological system that is hardwired to promote recall of auditory material. It seems that one would either have to assume yet another process or store specifically to account for a visual mid-list advantage, or to seek alternative explanations that might also be capable of explaining a recency advantage in sequences without an auditory component, like silently mouthed or lip-read lists.

Beaman [17] offers an explanation for the inverted modality effect, based primarily on data from a nonstandard, split-list, serial recall setting. In this study, participants had to reproduce a list in serial order but start with the last few items. It was speculated that "with visual presentation participants rely upon a visual code that supports recall of early items when recall of those items is delayed" [17, p. 387], implying that the visual superiority in pre-recency was peculiar to the split-list design. However, although the inverted modality effect has been observed in several split-list recall studies [17, 28], as noted, it is also observed in strict forward serial recall (e.g., [42, 62]).

In sum, it appears that the inverted modality effect is real and robust. Trivial explanations of the effect, like a concurrent articulation-like impact of late-list items on early items in a vocalized list, or visual advantages tied specifically to a split-list experimental design, are too restricted to account for all instances of the effect. Nevertheless, an explanation is clearly needed for how pre-recency performance on visual lists can match recency performance on auditory lists. Turning to the three major verbal short-term memory accounts that have been the subject of this chapter, it seems that the interference-based approach [67, 68] is at a loss for an explanation for the inverted modality effect. From the interference-based perspective, auditory items are represented in modality-dependent features that are not prone to interference from internal processes, and visual items are represented in modality independent internal features that can be interfered with by processes like inner speech. This explains the traditional modality effect: Visual items in recency are interfered with by internal processes, while the memory trace for auditory items remains largely unaffected [67]. Given these premises, it is difficult to conceive, however, how a mid-list visual advantage could arise. One might suggest that visual lists are encoded in terms of modality-dependent interference-resistant visual features up until mid-list. This suggestion is echoed by store-based explanations that have been proposed to account for the inverted modality effect. Thus it has been argued that participants sometimes opportunistically recruit additional visual codes, to be stored presumably in the visuospatial sketchpad [7, 11], to assist with visual list maintenance (see [12]). This would explain how performance on visual items in pre-recency can match high performance in recency on auditory items, even though according to the store-based perspective these have direct access to an otherwise modality-neutral phonological store [8], or a dedicated Precategorical Acoustic Store [30]. Kozlov ([55], Experiment 5) assessed the suggestion that the inverted modality effect occurs because of a reliance on visual codes [12, 17]. If such codes can indeed be recruited strategically to improve performance, then it seems reasonable to expect that they will be recruited whenever they are available. This was tested by contrasting forward serial recall of three types of lists: Auditory lists, visual lists that were silently read ("visual-silent"), and visual lists that had to be vocalized ("visual-vocalized"). The results reveal a clear inverted modality effect: Mid-list performance on visual-silent lists was higher than on visual-vocalized or on auditory lists. Indeed this pre-recency advantage of visual-silent lists matched the high-recency performance of auditory and visual-vocalized lists. These data confirm that the split-list report method is certainly not a precondition for the inverted modality effect [17], although, it remains possible that having to restructure the list accentuates the effect [17, 28]. Moreover, the current results challenge the suggestion that recall for visual-silent lists in pre-recency is superior to auditory due to the recruitment of additional visual codes [12, 17]. In the visual-vocalized condition, visual information was also available, and could therefore presumably have been recruited to increase performance in pre-recency. Instead, pre-recency performance on visual-vocalized lists was in fact poorer than in any other condition, undermining the idea that additional visual codes can be recruited at will to boost performance.

However, defenders of the visual code-recruitment account could counter that visual codes were recruited after all in the visual-vocalized condition: Performance in pre-recency in that condition may not have been comparable to that in the visual-silent condition because the need to articulate each item created a concurrent articulation-like effect (cf. [34, 72]). That is, any performance boost that recruitment of visual codes to maintain items in pre-recency offered was superseded by the damaging impact of having to vocalize the to-be-remembered list items. The observation that recall of items in recency in the visual-vocalized condition was as high as in the auditory condition could then be explained with the acoustic trace of the last item being preserved in an additional store such as the Precategorical Acoustic Store [30]. Alternatively, the last acoustic item could have been protected from interference because it was maintained in a special acoustic code [67, 73]. Kozlov ([55], Experiment 6) addressed this counterargument, adding to the conditions of Experiment 5 a condition requiring the silent mouthing of visually presented lists (visual-mouthed). The results show that pre-recency performance on visual-silent lists is higher than on any other presentation type in this experiment. Thus, the inverted modality effect was once more demonstrated. Further, as expected, performance on auditory and visual-vocalized lists in the present experiment was high in recency producing a U-shaped performance pattern overall. The same pattern of performance was also observed for the visual-mouthed lists, even though these lists were devoid of an auditory component.

There are several ways in which the present findings challenge the notion of a hardwired memory advantage for auditory material as propagated by the majority of verbal short-term memory theories. First, they demonstrate that one of the main problems for that notion, namely, the inverted modality effect—the observation of a pre-recency visual advantage that matches the auditory advantage in recency—is robust. Moreover, it cannot be explained away, as has been attempted at least from the store-based perspective (c.f. [12]), with participants opportunistically encoding to-be-remembered visual items in a visual code to improve pre-recency performance. In both the visual-mouthed and the visual-vocalized condition, additional visual information was available, yet in both conditions performance was worse at pre-recency than in the auditory condition. It could be objected that this is because the performance boost in pre-recency gained from the recruitment of visual codes was superseded by a concurrent articulation-like effect of the respective oral activity (vocalizing or mouthing). However, this objection would carry with it the prediction that there is a greater chance of the inverted modality effect transpiring with visual-mouthed list. This is because mouthed concurrent articulation is significantly less disruptive than vocalized concurrent articulation [60]. Contrary to this prediction, pre-recency performance in the visual-mouthed condition was lower than in the visual-vocalized condition, making the idea that additional visual codes are recruited to improve pre-recency performance even less likely.

The present results also raise the question of how relative recency performance can be equal on lists that consist of auditory input and lists that do not if there are additional stores [30], codes [73], or item feature representations [67] that selectively benefit auditory recency performance. Of course, one could argue that there are yet further codes [73] in which the mouthed list is maintained in an auditory-like fashion. However, having discarded additional visual codes as an explanation for the visual advantage in pre-recency, one has to wonder how useful it is in general to assume additional dedicated mechanisms or storage codes to explain superior performance on one task or modality or the other.

An alternative, more parsimonious, approach would be to appeal to a single construct that might explain the similar shapes of the auditory, vocalized, and mouthed serial position curves. According to the perceptual-gestural account [46, 47, 50, 51], such a construct might be perceptual organization. In the auditory domain, the perceptual system is known to organize sound into discrete auditory objects or streams based on physical characteristics of the different auditory inputs [21], according to Gestalt principles similar to those operating in the visual domain [54]. For example, if an auditorily presented list is followed by silence, then this creates a clear figure–ground contrast that defines the list boundary. This boundary, in turn, serves as an edge for the temporally extended acoustic object, and thus acts as an anchor that serves to disambiguate the order of items at the end of the auditory list (see [69]).

While such an instantiation of the perceptual-gestural view might also seem to suggest that high performance in recency is reserved for lists with an acoustic component, it is easy to see how salient list-edges might also be present outside of the acoustic domain. Indeed, this idea is at the core of the remaining experiments reported in this chapter. For example, if each list item has to be accompanied (due to instructions) by an articulatory gesture—as in some of the conditions of Experiments 5 and 6 [55]—this list-processing-relevant activity may in effect transform the succession of to-be-remembered items into a discrete temporally extended object, the beginning and end of which is defined by the onset and cessation of that activity (see also [60], for a similar argument). The figure–ground contrast between the presence and absence of the salient activity can arguably result in as strong an anchor for item order as the silence at the end of an acoustic list. Moreover, if the processing of the visual-silent list is particularly engaging such as, for example, when the letter list is not presented as a sequence of alpha numeric characters, but instead a sequence of visually presented silent gestures that the participant has to reconstruct into letter representations through lip-reading, then again the beginning and end of the (participant's) list-processing activity will create clear boundaries, thus accounting for the strong "auditory-like" recency found for lip-read lists [24, 32, 40].

The perceptual-gestural account seems to offer a promising framework, therefore, for developing a unified explanation for high-recency performance for auditory, visual-vocalized, visual-mouthed, and even lip-read lists, without having to invoke additional stores or codes. Instead, the simple assumption that the list-processing activity/cessation creates a figure–ground contrast that acts as an anchor for item order recall seems sufficient to explain high-recency performance for a range of conditions. However, it is still unclear why pre-recency performance on visual-silent lists should be higher than on auditory lists and indeed why this pre-recency performance should match the boost in performance that the auditory list receives from the order disambiguating silence at the end of the list. That is, how does the visual superiority in pre-recency (the inverted modality effect) arise given that there is no apparent anchor in the middle of the visual list?

One possibility is that visual pre-recency performance is superior because of a greater flexibility in subjectively restructuring a visual compared to an auditory to-be-remembered list. There is some indication that the inverted modality effect is more apparent when the to-be-remembered list has to be restructured, such as when lists have to be recalled in a split-list fashion, recalling the later list items first [17, 28]. Subjective restructuring also has an important role in the perceptual-gestural account. In typical serial recall tasks, such as the ones described in this chapter, the main challenge lies in the maintenance of order of a sequence of verbal items that are unconstrained by grammar or syntax. According to the perceptual-gestural view, there are several ways to meet this challenge. For example, it is possible, utilizing motor planning skills, to subjectively impose a prosodic rhythm onto the sequence [46], thus grouping the to-be-remembered sequence into smaller chunks, which has a clear benefit on performance (see e.g. [35]). Yet in order to use motor planning skills to upload the to-be-remembered list onto a subjective prosodic rhythm that perhaps would even impose grouping constraints, the list needs to be unconstrained perceptually, like visual-silent lists. Any perceived structure might be at odds with the subjective motor planning strategy. An example of this would be lists that are perceived automatically as temporally extended objects, like auditory lists. If the to-be-remembered list is encoded as an object, then the objecthood itself generates order cues for list items, particularly at the object boundaries, that is, the beginning and end of the list. Yet, since the very idea of an "object" denotes a cohesive and bound entity that is rigid and immutable [77], it seems probable that the list-object will be resilient to subjective motor planning-based strategies of imposing order. This could explain the traditional and the inverted modality effects: If an auditory to-be-remembered list is encoded as an object, memory will be particularly high at the list boundaries, but memory at mid-list will be higher for lists that, due to their perceptually unconstrained nature, can be easily restructured to fit a subjective motor plan. Kozlov ([55], Experiment 7) addressed this hypothesis by testing whether visual lists lend themselves better to restructuring than auditory lists. The results once more revealed an inverted modality effect: Strong auditory performance in recency matched by superior visual pre-recency performance. Whereas restructuring the list impaired overall performance, this was the case independently of modality. These results cast doubt on the hypothesis that visual performance in pre-recency is improved by the ability to flexibly fit the visual to-be-remembered list onto an articulatory plan, or subjectively group it. The notion that an auditory list is processed as a temporally extended perceptual object suggests a certain internal rigidity of the mental representation of the auditory list [77]. However, the current results suggest that an auditory list is no more internally cohesive than a visual list. Thus, even if fitting a to-be-remembered sequence onto a subjective prosodic rhythm were important for maintaining the order of the sequence, as postulated by the perceptual-gestural account [46, 47], the current experiment ([55], Experiment 7) does not offer evidence that this fitting process should be easier for visual silent-list than for auditory lists. The notion of greater flexibility of restructuring a visual list does not therefore seem to be an adequate explanation for the inverted modality effect.

Another motor planning skill that, according to the perceptual-gestural account [46], can be drawn upon when assembling a list of unconstrained verbal items into a recallable sequence is co-articulation. In a list that is unconstrained by grammar or semantics it might be possible to impose order by adjusting the articulation of one item to allow a smoother articulatory transition to the next [46, 65, 87]. Visual-silent lists may afford such a strategy more than auditory, visual-vocalized, or visual-mouthed lists. Hearing, vocalizing, or even mouthing a to-be-remembered item might activate a certain schema of how to articulate the item. Deviating from that schema strategically, in order to shape the item end through co-articulation as a cue for the next item to promote serial recall might thus be more difficult. Indeed, in the auditory domain there is much evidence for a tendency to articulatorily imitate even very subtle properties of a verbal utterance [38, 74, 79]. In contrast, with visual-silent lists participants are free to modify their articulation of the to-be-remembered items in order to facilitate co-articulatory transitions between items. Thus, particularly in parts of the serial position curve where item memory is not anchored at perceptual object edges, that is, in mid-list, visual-silent list performance could benefit. In order to establish whether articulatory factors underpin the visual pre-recency advantage Experiment 8 [55] examined whether it is reduced when co-articulatory planning of the list is impeded through concurrent articulation. Once more, an inverted modality effect was observed. Moreover, the size of the effect matched the size of the standard modality effect in recency, even when articulation was suppressed. The observation of the inverted modality effect under concurrent articulation makes it unlikely that the visual pre-recency advantage is dependent upon the ability to rehearse visual to-be-remembered items more effectively than auditory.

According to the perceptual-gestural account, in order to maintain the order of a sequence of verbal items that is unconstrained by semantics, grammar, or syntax, articulatory motor planning skills can be utilized. These are used to either fit the items onto a subjective prosodic rhythm or to modify the articulation of items to facilitate co-articulatory transitions between them [46]. Experiment 7 and 8 [55] have addressed several explanations as to how utilizing motor planning skills might be more efficient given visual-silent lists and how this might account for the inverted modality effect. Yet, the present results in particular make it seem unlikely that the inverted modality effect arises because of a greater ease in utilizing articulatory motor planning processes in the visual domain. If this were the case, then the inverted modality effect should be reduced under concurrent articulation. Instead, it was exacerbated. On the one hand, this clearly excludes gestural processes as the origin for the effect. On the other hand, looking toward perceptual processes for an explanation of the inverted modality effect also presents a conundrum: If auditory recency performance is improved by the order-disambiguating boundary that is provided by the silence at the end of a to-be-remembered list, how can performance on visual-silent lists in pre-recency match this improvement without an obvious mid-list boundary?

One possible difference between the various presentation types that might be the key to this paradox is the extent to which they constrain the obligatory encoding of the list in its entirety. This pertains to the idea of objects as cohesive entities [77]: If an item list is encoded, that processed beyond a mere sensory stage by the neurocognitive system, as temporally extended object, as seems to be the case with auditory visual-vocalized and visual-mouthed lists, then it stands to reason that all the elements of the object will also be encoded. On the other hand, if the to-be-remembered item list is perceived as a collection of discrete events, like with visual-silent lists, then there is no reason to assume that all the elements of the list will be encoded. Furthermore, if encoding only a subset of the presented to-be-remembered items would somehow benefit recall, then a presentation type that affords the freedom to selectively avoid encoding or "ignore" some of the items, that is, visual-silent presentation, should have a recall advantage over a presentation type in which each to-be-remembered item is encoded obligatorily. Naturally, this advantage would only extend to the items that are not being ignored, that is, items whose processing is not willfully prevented beyond a perceptual stage.

In order to demonstrate that the first of these assumptions, namely, that to-be-remembered lists that are represented as objects are indeed obligatorily processed in their entirety, a possible manipulation is to append an irrelevant, to-be ignored item, a suffix, at the end of a to-be-remembered list. If even a to-be-ignored suffix is obligatorily encoded as part of the to-be-remembered list, then all the to-be-remembered list items are likely to be subject to obligatory encoding, too. In point of fact, the detrimental effect of a suffix has been repeatedly demonstrated with auditory lists [40, 50, 51, 69] as well as vocalized and mouthed lists [40]. Specifically, it was found that if a suffix that shares perceptual properties with the to-be-remembered list items is appended at the end of the list then performance in recency is reduced. Moreover, it was shown that if the suffix is "captured," meaning that it is made to be perceived as part of an additional to-be-ignored stream of acoustic events, then the negative impact of the suffix on the to-be-remembered list diminishes [69]. Thus, it has been suggested that the suffix effect occurs because the suffix is obligatorily encoded as part of the to-be-remembered list that displaces the perceptual boundary at which the last to-be-remembered list item would otherwise be anchored. If, however, even a to-be-ignored suffix is obligatorily encoded as part of the to-be-remembered list, then it follows that every item in the to-be-remembered list will be obligatorily encoded, too. In contrast, previous research using visual-silent lists has constantly failed to find a suffix effect, even if the suffix shared (visual) perceptual properties with the other items of the to-be-remembered list, and trivial explanations like gaze aversion from the to-be-ignored stimulus have been ruled out [39]. Thus, even if the visual suffix is made to seem as part of the to-be-remembered list, there is little difficulty in ignoring it. This suggests that other visual list items might also be strategically ignored.

While past research on the suffix effect makes it seem likely that in lists perceived as temporally extended objects every item is obligatorily encoded, the data obtained in Experiment 9 [55] reveal a clear co-dependence between recall of the early and late items in auditory lists and a clear lack of such a dependence for visual-silent lists. Because these data demonstrate in an unprecedented way how temporally extended object lists are encoded in their entirety and lists not encoded as objects are not, the experiment is included here to bolster the argument that the inverted modality effect arises because participants can choose to ignore a part of a visually presented list. The inverted modality effect was once more replicated in this experiment. The superior performance on auditory lists in recency was accompanied by superior performance on visual lists in pre-recency. Moreover, in line with previous research [40, 50, 51, 69], the auditory suffix reduced auditory list performance in recency. If the suffix cannot be ignored, however, it would seem to follow that all the to-be-remembered items are also encoded.

A cluster analysis corroborates this conclusion. In the absence of the suffix, early and late list auditory items have a high likelihood of being co-recalled. This suggests that some obligatory holistic processing of the entire auditory list takes place, whereby whenever the early-list items are recalled well the last item is recalled well, too. This supports the notion of auditory list objecthood postulated by the perceptual-gestural account. The cluster analysis also shows that the suffix displaces the last to-be-remembered list item from the auditory list object edge. In conjunction, the cluster analysis makes it doubtful that any to-be-remembered items from the auditory list can be ignored. In contrast, cluster analysis of the visual lists reveals that there is a consistent trend to co-recall the first four visual items, with the late list items falling outside of this recall cluster. This, together with the comparatively low visual performance in recency, is in line with the idea that the visual list is not obligatorily encoded in its entirety.

In order to explain how the ability to strategically ignore portions of the visually presented to-be-remembered list might result in an encoding-based visual pre-recency advantage, it is necessary to assume that the short-term memory process is capacity limited and that overburdening the process beyond its capacity has a negative effect on recall. The idea of a capacity limit on short-term memory is widely accepted (see [8, 27, 64]). From the perceptual-gestural perspective [46, 47, 50, 51], in order to maintain the order of a sequence of dissociated verbal tokens, articulatory motor skills can be utilized to impose order, for example, by modifying item articulation to facilitate co-articulatory transitions. Yet, clearly, there must be a limit on the number of such co-articulatory modifications, and their complexity, that can be planned and maintained at any one time. When this number is exceeded, for example, as a result of encoding too many items, then accurate recall of order can no longer be guaranteed.

2.5 Conclusions

At the outset, this chapter introduced the "dragon" metaphor. The purpose of the metaphor was to illustrate how a well-defined single construct can easily explain a wide range of data, whilst still being improbable. At the same time, a belief in dragons constitutes a prime example of how inappropriate it is to believe in something that is nowhere to be found. The aim of this chapter is to show that the dragon metaphor is highly relevant for verbal short-term memory research. Thus, in the debate about the nature of verbal short-term memory, some (e.g., [8]) propose that it is accomplished by a dedicated mechanism, comprising a bespoke store that serves as a passive repository of decay-prone phonological memory traces, and an active articulatory control process to refresh these traces. Yet although the simple and elegant concept of a bespoke store has considerable explanatory power, it seems doubtful that a system dedicated exclusively to retaining phonological information for a brief period of time would have evolved. Indeed, the notion of the passive short-term store, because of its elegance and simultaneous unlikelihood, is rather reminiscent of the notion that travelers disappear on long journeys because of dragons.

Other theorists (e.g., [67]), rejecting the idea of a passive store in which information decays and an active refreshing mechanism, have argued that verbal short-term memory is an entirely passive process that is governed by retroactive interference. Yet, the processes that prominent interference-based models propose have not been identified with any neurological equivalent; they lurk somewhere within the neural architecture but only like a dragon behind a hill.

In this chapter, an attempt was made to advocate yet another alternative approach, considering short-term memory as an emergent property of general receptive and productive mechanisms, with speech-related productive mechanisms being employed particularly frequently on verbal short-term memory task due to the linguistic nature of the to-be-remembered material. This allows a clear mapping of processes associated with verbal short-term memory onto brain regions associated with perception, (speech-) planning, and production. At the same time, there is no necessity to invoke yet another dedicated system, thus violating evolutionary parsimony. Thus, the perceptual-gestural approach to verbal short-term memory could hold the key to solving the "dragon" problem. To confirm this, the empirical work presented in this chapter focused on demonstrating that the perceptual-gestural account offers an equal or better explanation of a range of data obtained from behavioral and neuroscientific studies, than either the store-based or the interference-based approach.

References

1. Acheson, D.J., MacDonald, M.C.: Verbal working memory and language production: common approaches to the serial ordering of verbal information. Psychol. Bull. **135**, 50–68 (2009)
2. Anderson, J.A.: An Introduction to Neural Networks. MIT Press, Cambridge (1995)
3. Atkins, W.B., Baddeley, A.D.: Working memory and distributed vocabulary learning. Appl Psycholinguist. **19**, 537–552 (1998)
4. Baddeley, A.D.: Working Memory. Oxford University Press, Oxford (1986)
5. Baddeley, A.D.: Is working memory working? The fifteenth Bartlett lecture. Q. J. Exp. Psychol. **44A**, 1–31 (1992)
6. Baddeley, A.D.: Human Memory: Theory and Practice, Rev. edn. Psychology Press, Hove (1997)
7. Baddeley, A.: The episodic buffer: a new component of working memory? Trends Cogn. Sci. **4**, 417–432 (2000)
8. Baddeley, A.D.: Working memory: looking back and looking forward. Nat. Rev. Neurosci. **4**, 829–839 (2003)
9. Baddeley, A.D.: Working memory, thought and action. Oxford University Press, Oxford (2007)
10. Baddeley, A.D., Gathercole, S.E., Papagno, C.: The phonological loop as a language learning device. Psychol. Rev. **105**, 158–173 (1998)
11. Baddeley, A.D., Hitch, G.: Working memory. In: Bower, G.H. (ed.) The Psychology of Learning and Motivation: Advances in Research and Theory, vol. 8, pp. 47–89. Academic, New York (1974)
12. Baddeley, A.D., Larsen, J.D.: The phonological loop unmasked? A comment on the evidence for a "perceptual-gestural" alternative. Q. J. Exp. Psychol. **60**, 497–504 (2007)
13. Baddeley, A., Lewis, V., Vallar, G.: Exploring the articulatory loop. Q. J. Exp. Psychol. **36**, 233–252 (1984)
14. Baddeley, A.D., Thomson, N., Buchanan, M.: Word length and the structure of short-term memory. J. Verbal Learn. Verbal Behav. **14**, 575–589 (1975)
15. Baddeley, A.D., Wilson, B.: Phonological coding and short-term memory in patients without speech. J. Mem. Lang. **24**, 490–502 (1985)
16. Baker, J.R., Bezance, J.B., Zellaby, E., Aggleton, J.P.: Chewing gum can produce context-dependent effects upon memory. Appetite. **43**, 207–210 (2004)
17. Beaman, C.P.: Inverting the modality effect in serial recall. Q. J. Exp. Psychol. **55**, 371–389 (2002)
18. Beaman, C.P., Jones, D.M.: The role of serial order in the irrelevant speech effect: tests of the changing state hypothesis. J. Exp. Psychol. Learn. Mem. Cogn. **23**, 459–471 (1997)
19. Beaman, C.P., Neath, I., Surprenant, A.M.: Phonological similarity effects without a phonological store: an individual differences model. In: McNamara, D.S., Trafton, J.G. (eds.) Proceedings of the 29th Annual Conference of the Cognitive Science Society, pp. 89–94. Cognitive Science Society, Austin (2007)
20. Bishop, D.V.M., Robson, J.: Unimpaired short-term memory and rhyme judgement in congenitally speechless individuals: implications for the notion of "articulatory coding". Q. J. Exp. Psychol. **41**, 123–140 (1989)
21. Bregman, A.S.: Auditory Scene Analysis: The Perceptual Organisation of Sound. MIT Press, Cambridge (1990)
22. Buchsbaum, B.R., D'Esposito, M.: The search for the phonological store: from loop to convolution. J. Cogn. Neurosci. **20**, 762–778 (2008)
23. Buschke, H.: Relative retention in immediate memory determined by the missing scan method. Nature. **200**, 1129–1130 (1963)
24. Campbell, R., Dodd, B.: Hearing by eye. Q. J. Exp. Psychol. **32**, 85–99 (1980)
25. Conrad, R.: Acoustic confusions in immediate memory. Br. J. Psychol. **55**, 75–84 (1964)

26. Conrad, R., Hull, A.J.: Input modality and the serial position curve in short-term memory. Psychonomic Sci. **10**, 135–136 (1968)
27. Cowan, N.: The magical number 4 in short-term memory: a reconsideration of mental storage capacity. Behav. Brain Sci. **24**, 87–114 (2001)
28. Cowan, N., Saults, J.S., Brown, G.D.A.: On the auditory modality superiority effect in serial recall: separating input and output factors. J. Exp. Psychol. Learn. Mem. Cogn. **30**, 639–644 (2004)
29. Crowder, R.G.: The role of one's own voice in immediate memory. Cogn. Psychol. **1**, 157–178 (1970)
30. Crowder, R.G., Morton, J.: Precategorical acoustic storage (PAS). Percept. Psychophys. **5**, 365–373 (1969)
31. Davis, C., Kleinman, J.T., Newhart, M., Gingis, L., Pawlak, M., Hillis, A.E.: Speech and language functions that require a functioning Broca's area. Brain Lang. **105**, 50–58 (2008)
32. de Gelder, B., Vroomen, J.: Abstract versus modality-specific memory representations in processing auditory and visual speech. Mem. Cognit. **20**, 533–538 (1992)
33. Ellis, A.W.: Errors in speech and short-term memory: the effects of phonemic similarity and syllable position. J. Verbal Learn. Verbal Behav. **19**, 624–634 (1980)
34. Ellis, N.R.: Evidence for two storage processes in short-term memory. J. Exp. Psychol. **50**, 390–391 (1969)
35. Frankish, C.: Perceptual organization and precategorical acoustic storage. J. Exp. Psychol.: Learn. Mem. Cogn. **15**, 469–479 (1989)
36. Freeman, G.L.: Dr. Hollingworth on chewing as a technique of relaxation. Psychol. Rev. **47**, 491–493 (1940)
37. Glenberg, A.M., Swanson, N.G.: A temporal distinctiveness theory of recency and modality effects. J. Exp. Psychol. Learn. Mem. Cogn. **12**, 3–15 (1986)
38. Goldinger, S.D.: Echoes of echoes? An episodic theory of lexical access. Psychol. Rev. **105**, 251–279 (1998)
39. Greene, R.L.: Stimulus suffixes and visual presentation. Mem. Cognit. **5**, 497–503 (1987)
40. Greene, R.L., Crowder, R.G.: Modality and suffix effects in the absence of auditory stimulation. J. Verbal Learn. Verbal Behav. **23**, 371–382 (1984)
41. Guerard, K., Jalbert, A., Neath, I., Surprenant, A.M., Bireta, T.J.: Irrelevant tapping and the acoustic confusion effect: the effect of spatial complexity. Exp. Psychol. **56**, 367–374 (2009)
42. Harvey, A.J., Beaman, C.P.: Input and output modality effects in immediate serial recall. Memory. **15**, 693–700 (2007)
43. Henson, R.N.A.: Short-term memory for serial order: the start-end model. Cogn. Psychol. **36**, 73–137 (1998)
44. Hickok, G.: The functional neuroanatomy of language. Phys. Life Rev. **6**, 121–143 (2009)
45. Huang, Y., Edwards, M.J., Rounis, E., Bhatia, K.P., Rothwell, J.C.: Theta burst stimulation of the human motor cortex. Neuron. **45**, 201–206 (2005)
46. Hughes, R.W., Marsh, J.E., Jones, D.M.: Perceptual–gestural (mis)mapping in serial short-term memory: the impact of talker variability. J. Exp. Psychol. Learn. Mem. Cogn. **35**, 1411–1425 (2009)
47. Hughes, R.W., Marsh, J.E., Jones, D.M.: Role of serial order in the impact of talker variability in short-term memory: testing a perceptual organization-based account. Mem. Cognit. **39**, 1435–1447 (2011)
48. James, W.: The Principles of Psychology. Holt, Rinehart & Winston, New York (1890)
49. Johnson, A.J., Miles, C.: Chewing gum and context-dependent memory: the independent roles of chewing gum and mint flavour. Br. J. Psychol. **99**, 293–306 (2008)
50. Jones, D.M., Hughes, R.W., Macken, W.J.: Perceptual organization masquerading as phonological storage: further support for a perceptual-gestural view of short-term memory. J. Mem. Lang. **54**, 265–281 (2006)
51. Jones, D.M., Macken, W.J., Nicholls, A.P.: The phonological store of working memory: is it phonological and is it a store? J. Exp. Psychol. Learn. Mem. Cogn. **30**, 656–674 (2004)

52. Jones, D.M., Tremblay, S.: Interference in memory by process or content? A reply to Neath (2000). Psychon. Bull. Rev. **7**, 550–558 (2000)
53. Kirschen, M.P., Davis-Ratner, M.S., Jerde, T.E., Schraedley-Desmond, P., Desmond, J.E.: Enhancement of phonological memory following transcranial magnetic stimulation (TMS). Behav. Neurol. **17**, 187–194 (2006)
54. Koffka, K.: Principles of Gestalt Psychology. Harcourt, Brace & World, New York (1935)
55. Kozlov, M.D.: Verbal short-term memory: Cognitive and neuroscientific tests of a perceptual-gestural account (Doctoral dissertation). Cardiff University, School of Psychology (2012)
56. Kozlov, M.D., Hughes, R.W., Jones, D.M.: Gummed up memory: chewing gum impairs short-term recall. Q. J. Exp. Psychol. **65**, 501–513 (2011)
57. Lashley, K.S.: The problem of serial order in behavior. In: Jeffress, L.A. (ed.) Cerebral Mechanisms in Behavior. Wiley, New York (1951)
58. LeCompte, D.C.: Irrelevant speech, serial rehearsal and temporal distinctiveness: a new approach to the irrelevant speech effect. J. Exp. Psychol. Learn. Mem. Cogn. **22**, 1154–1165 (1996)
59. Logie, R.H., Della Sala, S., Wynn, V., Baddeley, A.D.: Visual similarity effects in immediate serial recall. Q. J. Exp. Psychol. **53**, 626–646 (2000)
60. Macken, W.J., Jones, D.M.: Functional characteristics of the inner voice and the inner ear: single or double agency? J. Exp. Psychol. Learn. Mem. Cogn. **21**, 436–448 (1995)
61. Macken, B., Taylor, J.C., Kozlov, M.D., Hughes, R.W., Jones, D.M.: Memory as embodiment: the case of modality and serial short-term memory. Cognition. **155**, 113–124 (2016)
62. Maylor, E.A., Vousden, J.I., Brown, G.D.A.: Adult age differences in short-term memory for serial order: Data and a model. Psychol. Aging. **14**, 572–594 (1999)
63. Maizey, L., Allen, C.P.G., Dervinis, M., Verbruggen, F., Varnava, A., Kozlov, M., Adams, R.C., Stokes, M., Klemen, J., Bungert, A., Hounsell, C.A., Chambers, C.D.: Comparative incidence rates of mild adverse effects to transcranial magnetic stimulation. Clin. Neurophysiol. **124**, 536–544 (2013)
64. Miller, G.A.: The magical number seven, plus or minus two: some limits on our capacity for processing information. Psychol. Rev. **63**, 81–97 (1956)
65. Murray, A., Jones, D.M.: Articulatory complexity at item boundaries in serial recall: the case of Welsh and English digit span. J. Exp. Psychol. Learn. Mem. Cogn. **28**, 594–598 (2002)
66. Murray, D.J.: Articulation and acoustic confusability in short-term memory. J. Exp. Psychol. **78**, 679–684 (1968)
67. Nairne, J.S.: A feature model of immediate memory. Mem. Cognit. **18**, 251–269 (1990)
68. Neath, I.: Modeling the effects of irrelevant speech on memory. Psychon. Bull. Rev. **7**, 403–423 (2000)
69. Nicholls, A.P., Jones, D.M.: Capturing the suffix: cognitive streaming in immediate serial recall. J. Exp. Psychol. Learn. Mem. Cogn. **28**, 12–28 (2002)
70. Ogar, J., Slama, H., Dronkers, N., Amici, S., Gorno-Tempini, M.L.: Apraxia of speech: an overview. Neurocase. **11**, 427–432 (2005)
71. Paulesu, E., Frith, C.D., Frackowiak, R.S.J.: The neural correlates of the verbal component of working memory. Nature. **362**, 342–345 (1993)
72. Penney, C.G.: Modality effects in short-term verbal memory. Psychol. Bull. **82**, 68–84 (1975)
73. Penney, C.G.: Modality effects and the structure of short-term verbal memory. Mem. Cognit. **17**, 398–422 (1989)
74. Pickering, M.J., Garrod, S.: Do people use language production to make predictions during comprehension? Trends Cogn. Sci. **11**, 105–110 (2007)
75. Romero, L., Walsh, V., Papagno, C.: The neural correlates of phonological short-term memory: a repetitive transcranial magnetic stimulation study. J. Cogn. Neurosci. **18**, 1147–1155 (2006)
76. Saito, S.: What effect can rhythmic finger tapping have on the phonological similarity effect? Mem. Cognit. **22**, 181–187 (1994)
77. Spelke, E.S.: Principles of object perception. Cognit. Sci. **14**, 29–56 (1990)
78. Stephens, R., Tunney, R.J.: Role of glucose in chewing gum-related facilitation of cognitive function. Appetite. **43**, 211–213 (2004)

79. Street Jr., R.L., Cappella, J.N.: Social and linguistic factors influencing adaptation in children's speech. J. Psycholinguist. Res. **18**, 497–519 (1989)
80. Tremblay, S., Parmentier, F.B.R., Guérard, K., Nicholls, A.P., Jones, D.M.: A spatial modality effect in serial memory. J. Exp. Psychol. Learn. Mem. Cogn. **32**, 1208–1215 (2006)
81. Vallar, G.: Memory systems: the case of phonological short-term memory. A festschrift for cognitive neuropsychology. Cogn. Neuropsychol. **23**, 135–155 (2006)
82. Vallar, G., Baddeley, A.D.: Fractionation of working memory: neuropsychological evidence for a phonological short-term store. J. Verbal Learn. Verbal Behav. **23**, 151–161 (1984)
83. Verbruggen, F., Aron, A.R., Stevens, M.A., Chambers, C.D.: Theta burst stimulation dissociates attention and action updating in human inferior frontal cortex. Proc. Natl. Acad. Sci. USA. **107**, 13966–139671 (2010)
84. Vishwanathan, A., Bi, G.Q., Zeringue, H.C.: Ring-shaped neuronal networks: a platform to study persistent activity. Lab Chip. **11**, 1081–1088 (2011)
85. Waters, G.S., Rochon, E., Caplan, D.: Role of high level speech planning in rehearsal: evidence from patients with apraxia of speech. J. Mem. Lang. **31**, 54–73 (1992)
86. Wilkinson, L., Scholey, A., Wesnes, K.: Chewing gum selectively improves aspects of memory in healthy volunteers. Appetite. **38**, 235–236 (2002)
87. Woodward, A.J., Macken, W.J., Jones, D.M.: Linguistic familiarity in short-term memory: a role for (co-) articulatory fluency? J. Mem. Lang. **58**, 48–65 (2008)

Chapter 3
In-Memory Computing: The Integration of Storage and Processing

Saeideh Shirinzadeh and Rolf Drechsler

Abstract In today's computer architectures, the data is processed and stored in two separate units. Therefore, a communication system is needed to transfer the data between these two main components. The required time for communcation to access the stored data limits performance of the processor to much less than its potential. This problem is especially challenging due to the emergence of new applications which require huge amount of data to be processed in real time. Some of the advanced emerging memory technologies may provide a solution for this problem by allowing to perform logic operations within the memory cells. This enables a new generation of computer architectures in which the memory and the processor are unified. In this context, this chapter presents the state of the art for in-memory computation of logical functions using the *Resistive RAM* technology. In particular, the chapter introduces a logic-in-memory computer architecture and shows how it can be programmed to efficiently compute arbitrary arithmetic and control functions such as those frequently used in modern computers' processing units.

3.1 Introduction

The present-day computer architecture was named after John von Neumann who had a great influence on the way how the computers are designed today. The von Neumann architecture, in a nutshell, consists of three components: the storage to maintain the raw data and the processed information, the processor to perform calculations, and the bus as the communication system transferring the data between the two other components. Among the three mentioned components, the storage

S. Shirinzadeh (✉) · R. Drechsler
Group of Computer Architecture, Faculty 3 - Mathematics and Computer Science, University of Bremen, Bremen, Germany

Cyber-Physical Systems, DFKI GmbH, Bremen, Germany
e-mail: s.shirinzadeh@uni-bremen.de; drechsler@uni-bremen.de

C. S. Große, R. Drechsler (eds.), *Information Storage*,
https://doi.org/10.1007/978-3-030-19262-4_3

causes the largest latency which is referred to as memory bottleneck since it limits the overall performance. In other words, the processor cannot work at a higher speed than the memory due to the required waiting time for the data to be retrieved.

The storage of a computer consists of two parts: the permanent storage, the so-called hard disk, and the temporary storage called *Random-Access Memory* (RAM) which is also shortly referred as memory. The RAM only retains information as long as it is supplied by electrical power, while the disk keeps its contents even if the computer is switched off. Among these two parts of the storage system, what is blamed for the high latency is the disk and not the RAM, as it performs up to 20 times faster. To clarify this better, let us consider a typical personal computer system with a 64-bit[1] quad-core processor. This means that each of the four independent cores is capable of manipulating data as big as 64 bits 4 billion times[2] per second. Such a computer system with an average price in the market possibly possesses a RAM of size 4 billion bytes[3] and a disk of much greater size equal to 500 billion bytes. In this condition, only a single core can process an amount of data which may potentially need larger memory than what is provided by RAM. Consequently, the rest of the data under process needs to be managed in disk.

Now, one can think that the memory bottleneck can be simply solved by allocating large amounts of RAM enough to be used as the only storage space when communication is necessary with the processor and using the disk only for archiving when the resulted information need to be stored permanently. Nevertheless, the amount of disk in any typical computer is considerably larger than RAM due to its high price. This makes it costly to manage all the data within RAM as explained in the scenario above.

The emerging applications such as the *Internet of Things* (IoT) and *big data* deal with big amounts of data which should be processed fast. Accordingly, these applications need much faster interaction between the storage and the processing units. For example, the popular Internet companies such as Google, Facebook, Amazon, and Twitter with a wide spread presence in our daily life deal with huge amount of data which needs to be processed in real time. Today, the requirements of emerging applications and the advancements in hardware have resulted in a variety of *in-memory data management systems* which keep the data in the RAM all the time to speed up the communication between storage and processor. *Resistive RAM* (RRAM) is a promising technology for such in-memory storage systems [8, 20–22, 24]. RRAM is a *nonvolatile memory* (NVM) technology, which means that it does not need power supply to keep its contents similar to disk. However, RRAM provides considerably better performance compared to disk and other NVM technologies such as the widely used flash memories.

---

[1] The smallest unit of measurement used to quantify computer data which can adopt either value 0 or 1.

[2] It is equal to 4 GHz (4×10^9) clock rate, that is, the speed of the processor to execute tasks.

[3] A byte consists of 8 bits.

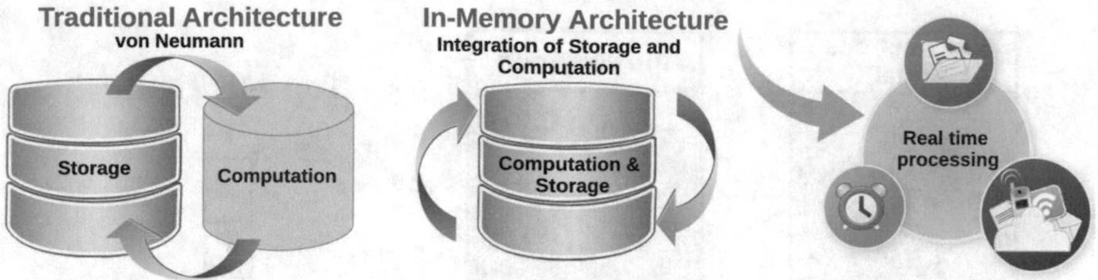

Fig. 3.1 von Neumann vs. in-memory computer architecture

The interesting properties of RRAM not only allow to improve the data transfer between the storage and the processor but also provide the opportunity to abolish the need for storage-processor communication in the von Neumann architecture. An RRAM device stores information as internal resistance which can be switched between two high and low values designating the binary states. The resistive switching property of RRAM enables to perform primitive logic operations which are sufficient to express any logical function directly within the memory. Indeed, RRAM can be programmed to calculate any task performed within current computer processors. Furthermore, the RRAM is automatically updated by the results of the calculation without additional latency to transfer and write the information which is inevitable when the memory and processor are separated. This integration of memory and processor, that is, *in-memory computing*, enables a new generation of computers beyond the limits of the traditional von Neumann architecture (see Fig. 3.1).

This chapter presents state-of-the-art approaches for in-memory computing within RRAM arrays. The chapter surveys in-memory computing in its two aspects. In the first part, the fundamental aspect of synthesis of logical functions as performed by computer processors is presented for in-memory computing according to the state of the art. The second part presents a simple in-memory computer architecture.

The rest of this chapter is organized as the following. In Sect. 3.2, a brief introduction to RRAM technology and its switching behavior is provided. Section 3.3 discusses the central question, that is, design for in-memory computing. The section presents a graph-based approach using three different representations. In Sect. 3.4, a generic state-of-the-art logic-in-memory architecture is introduced and its design issues are discussed. The concluding remarks of the chapter are presented in Sect. 3.5.

3.2 RRAM: A Technology for Memory and Computation

An RRAM cell is a two-terminal device which is made of a metal oxide sandwiched by two metal electrodes. The RRAM basically performs like an electrical resistor which internal resistance allocates two stable low and high values designating the

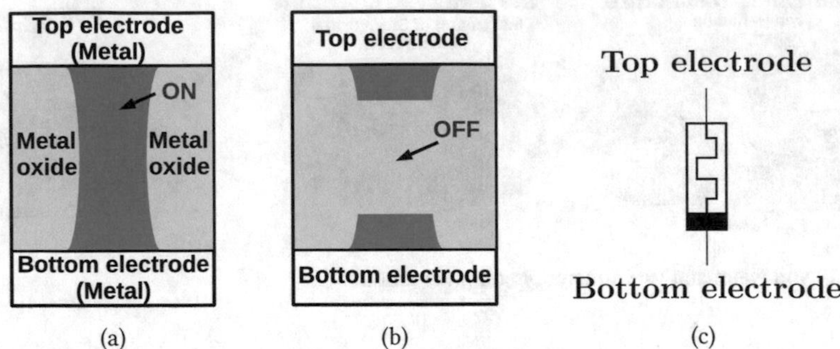

Fig. 3.2 Structure of an RRAM device: (**a**) low-resistance state (on), (**b**) high-resistance state (off). (**c**) Symbol of an RRAM device

binary values 1 and 0, respectively. The switching between the alternate states is possible by applying appropriate voltages to the device which can be exploited to execute logic operations within RRAM devices. Figure 3.2 shows the structure of an RRAM device in its two different resistance states which correspond to on and off switches. As the figure shows, the device is on when it is in low-resistance state or in other words the electrical conductance is high and vice versa. In the following, it is explained how two fundamental operations can be executed within an RRAM device to be used as the basis for design of RRAM-based logical circuits.

3.2.1 *Material Implication* (IMP)

Material Implication (IMP) and FALSE operation, that is, assigning the output to logic 0, are sufficient to express any Boolean function [4]. Figure 3.3a shows the circuit implementation of an IMP gate which was proposed in [4]. P and Q denote two RRAM devices connected to a load resistor R_G. Three voltage levels V_{SET}, V_{COND}, and V_{CLEAR} are applied to the devices to execute IMP and FALSE operations by switching between low-resistance (logic 1) or high-resistance (logic 0)

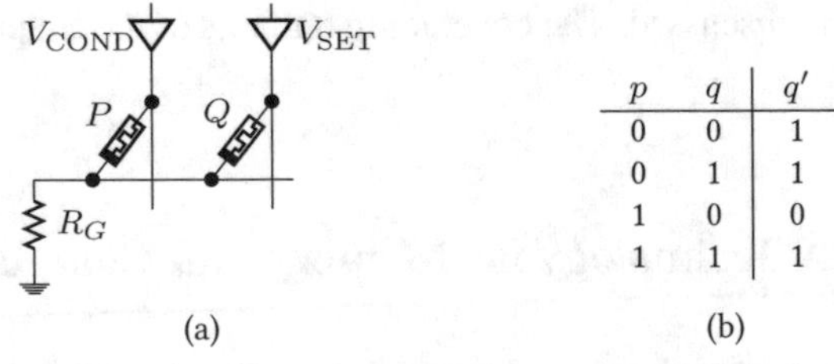

p	q	q'
0	0	1
0	1	1
1	0	0
1	1	1

Fig. 3.3 Material implication (IMP) operation. (**a**) Implementation of IMP using RRAM devices. (**b**) Truth table for IMP ($q' \leftarrow p$ IMP $q = \overline{p} + q$) [4]

states. The resistance state of an RRAM device is switched to 0 and 1 by applying V_{CLEAR} and V_{SET} to its voltage driver, respectively. To execute IMP, two voltage levels V_{SET} and V_{COND} are applied to the switches P and Q simultaneously. The interaction of the two RRAM devices under the aforementioned voltages executes the IMP operation.

The IMP operation changes the state of one of the switches, that is, Q in Fig. 3.3a, which is called the *work device*. The other RRAM device, that is, P in Fig. 3.3a, keeps its initially set value after the execution of operation and is called *input device*. The corresponding changes caused by the IMP operation in the state of the switches P and Q are shown in Fig. 3.3b. The IMP operation performs according to the current states of both switches such that switch Q is set to 1 if $p = 0$ and it retains its current state if $p = 1$ [4].

3.2.2 *Built-In Majority Operation* (RM₃)

In [7], it was shown that RRAM devices can implement an intrinsic majority behavior. More precisely, the resistive switching property of a single RRAM device works as a majority function of three variables combined with inversion, that is, assigning the logical complement of a variable $\bar{x}$. The three-input majority function $M(x, y, z) = x \cdot y + x \cdot z + y \cdot z$ is a logical operator which returns the output equal to the logical value of 2 or 3 of its input variables, for example, the output is 1 if at least two of the inputs are equal to 1. The combination of majority and inversion natively implemented within RRAM provides a universal set of primitive operations which can be used to construct all logical functions and therefore is of high interest for logic-in-memory computing.

Denoting the top and bottom terminals of an RRAM device by P and Q, the memory can be switched with a negative or positive voltage V_{PQ} based on the device polarity. Here, we assume that the logic statements $(P = 1, Q = 0)$ switch the RRAM device to logic 1, $(P = 0, Q = 1)$ switch the device to logic 0, and $(P = Q)$ do not change the current state of the device. Accordingly, we can make the tables shown in Fig. 3.4 for the next state of the switch (R') when the current state (R) is either 0 or 1. In the following, the Boolean relations represented by tables in Fig. 3.4 are extended which formally express the built-in majority operation of RRAM devices [7]:

$$\begin{aligned}
R' &= (P \cdot \overline{Q}) \cdot \overline{R} + (P + \overline{Q}) \cdot R \\
&= P \cdot R + \overline{Q} \cdot R + P \cdot \overline{Q} \cdot \overline{R} \\
&= P \cdot R + \overline{Q} \cdot R + P \cdot \overline{Q} \cdot R + P \cdot \overline{Q} \cdot \overline{R} \\
&= P \cdot R + \overline{Q} \cdot R + P \cdot \overline{Q} \\
&= M(P, \overline{Q}, R)
\end{aligned}$$

P	Q	R	R'
0	0	0	0
0	1	0	0
1	0	0	1
1	1	0	0

$R' = P \cdot \overline{Q}$

P	Q	R	R'
0	0	1	1
0	1	1	0
1	0	1	1
1	1	1	1

$R' = P + \overline{Q}$

$$\mathbf{RM_3(P, Q, R)} : R' = (P \cdot \overline{Q}) \cdot \overline{R} + (P + \overline{Q}) \cdot R = M(P, \overline{Q}, R)$$

Fig. 3.4 The intrinsic majority operation within an RRAM device [7]

The equation obtained above shows that the next state of an RRAM device is equal to the output of a three-input majority gate which inputs are the current state of the device, the logical state of the top electrode, and the inverted logical state of the bottom electrode. This operation is referred to three-input resistive majority RM3 which RM3(x, y, z), such as RM3$(x, y, z) = M(x, \bar{y}, z)$ [7]. According to RM3, the next state of a resistive switch is equal to a result of a built-in majority gate when one of the three variables x, y, and z is already preloaded and the variable corresponding to the logic state of the bottom electrode is inverted.

3.3 Central Question: Design for In-Memory Computing

As explained before, the basic operations provided by the resistive switching property of RRAM enable to realize any arbitrary logical function with a sequence of IMP or RM3 operations on a memory array. This brings us to the central question of this chapter which is "how to design for logical RRAM-based in-memory computing?" This section presents a comprehensive state-of-the-art approach for logic-in-memory computing based on the use of logic representations, that is, *Binary Decision Diagrams* (BDDs), *And-Inverter Graph* (AIG), and *Majority-Inverter Graph* (MIG). In the following, the employed representations are introduced and then the design methodology for each representation is explained.

3.3.1 Logic Representations

3.3.1.1 Binary Decision Diagrams (BDDs)

Binary Decision Diagram (BDD; e.g., [6]) is a graph-based structure that can provide a compact representation for most of the Boolean functions with practical size. BDD is derived from Shannon decomposition $f = x_i f_{x_i} \oplus \bar{x}_i f_{\bar{x}_i}$ [12]. Applying this decomposition recursively allows dividing the function into many smaller sub-

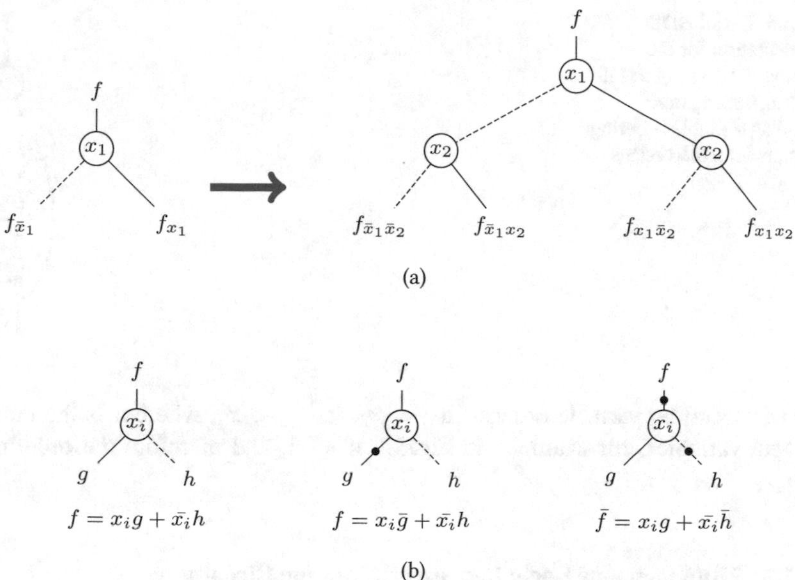

Fig. 3.5 (**a**) Applying Shannon decomposition recursively to find the BDD representation of function f. (**b**) Examples depicting the concepts of regular and complemented edges within BDDs

functions, which constitute the nodes of the BDD representation. Figure 3.5a shows this decomposition which results in BDD nodes with their sub-functions which are shown at the edge of node successors. Each node has two high and low successors denoted by solid and dashed lines, referring to assignments $x_i = 1$, and $x_i = 0$, respectively.

BDDs are compact due to the fact that for many functions of practical interest, smaller sub-functions occur repeatedly and need to be represented only once. Also, a sub-function and its complement may be represented by the same node which further reduces the number of nodes. The complement attribute is shown by a dot on the so-called complemented edge and should be interpreted as a logically inverted function at the edge (see Fig. 3.5b). An example of a BDD with complemented edges is shown in Fig. 3.6 which represents a function with four variables. The nodes corresponding to each input variable x_i represent a BDD level i which needs to be calculated in order, starting from the bottom of the graph to the root node f.

BDD is a powerful data structure for many applications. For the majority of applications using BDDs, the number of nodes in the BDD, also known as BDD's size, is the main cost metric. BDDs are canonical w.r.t. the variable ordering which also determines the size of the BDD. Improving the variable ordering for BDDs is computationally hard [3]. This has been a motivation behind many heuristic approaches aiming to find good orderings which result in smaller BDDs. Throughout this chapter, we consider initial BDD representations before optimization with

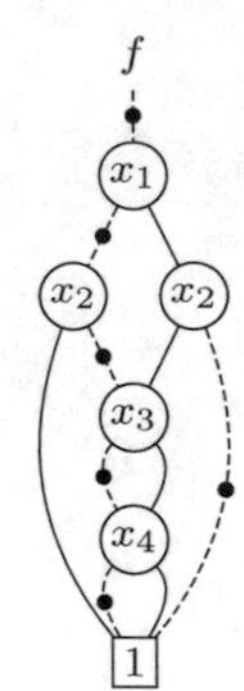

Fig. 3.6 Initial BDD representation for the function $f = (x_1 \oplus x_2) + (x_3 \oplus x_4)$, using the ascending variable ordering and complemented edges

a fixed ascending variable ordering $x_1 < x_2 < \cdots < x_n$, where n is the number of input variables, for example, in Fig. 3.6 $n = 4$, and therefore the ordering is $x_1 < x_2 < x_3 < x_4$.

3.3.1.2 Homogeneous Logic Representations for Circuits

In this chapter, the use of *And-Inverter Graphs* (AIGs [10]) and *Majority-Inverter Graphs* (MIGs [1]) is shown for logic-in-memory computing as homogeneous logic representations. Each node in both representations designates a basic logic operation, that is, $x \cdot y$ (AND also called conjunction) in case of AIGs, and $M(x, y, z) = x \cdot y + x \cdot z + y \cdot z$ (majority of three) in case of MIGs. Inverters are represented in terms of complemented edges, and regular edges represent non-complemented inputs.

An AND gate can be simply represented by a majority gate by adding a third constant input of value 0 (see Fig. 3.7). Accordingly, MIGs can be directly obtained from AIG representations by applying the node conversion over the entire graph. Logic representations in MIGs are at least as compact as in AIGs, since each AND node can be mapped to exactly one majority node; we have $x \cdot y = M(x, y, 0)$. However, even smaller MIGs can be obtained if their capability of compactness is fully exploited such that no node in the graph has constant inputs [2]. For this purpose, a Boolean algebra including primitive and advanced transformations was proposed in [1] in order to optimize MIGs. The transformations can be iteratively

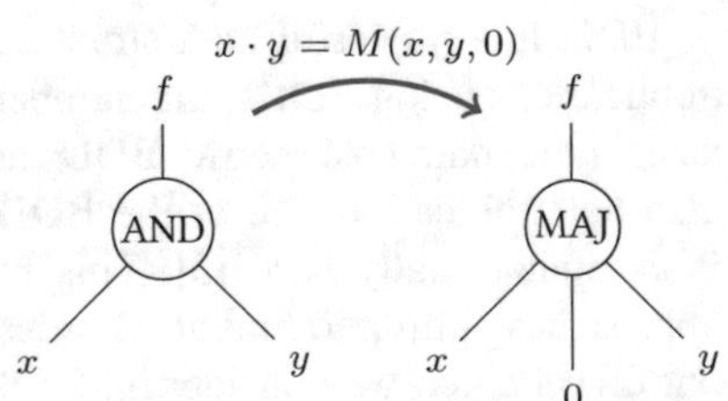

Fig. 3.7 A single-node AIG transposed to its equivalent MIG

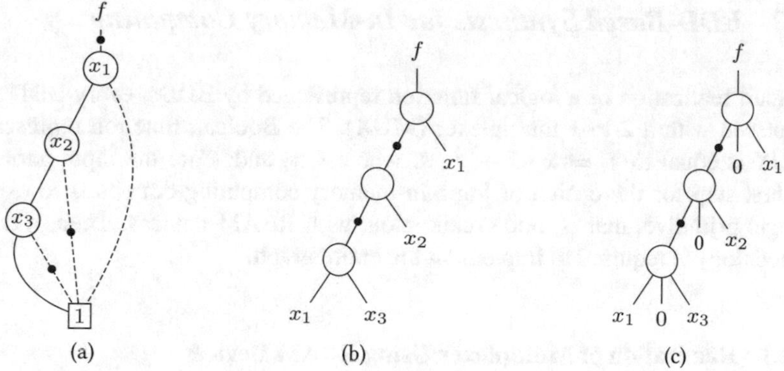

Fig. 3.8 Logic representations for an example function with three input variables. (**a**) BDD. (**b**) AIG. (**c**) MIG

applied to an existing MIG to get a more efficient representation. This way, any MIG can be transformed to another logically equivalent but more efficient MIG [2]. However, obtaining the desired MIG may be only possible with a long sequence of transformations which might be even impractical. In general, AIG and MIG both allow for efficient representation and thus are used as the major data structure in state-of-the-art logic synthesis tools.

Figure 3.8 shows all of the three discussed representations for a Boolean function with three input variables. All of the graphs have the same number of nodes and levels, that is, partitions with equal distance or computational latency from the root function f, which are equal to three for both metrics. The number of nodes and levels, also known as the size and depth, are considered important features of a representation because they determine the area and speed of the resulting circuits, respectively. In particular, the latency of the resulting circuits highly matters for in-memory computing. Therefore, optimization of the logic representations with respect to the number of levels of the graphs is very important. However, for BDDs the number of levels may only be optimized if one or more variables are entirely eliminated from the graph. Otherwise, the number of levels is constant and equal to the number of input variables of the presented function.

In the following, design for logic-in-memory computing using the introduced logic representation is explained. For each representation, circuit realizations of graph nodes with RRAM devices, design methodology to compute the graph on memory array, and optimization to lower latency and the number of devices are discussed in order.

3.3.2 BDD-Based Synthesis for In-Memory Computing

In circuit realization of a logical function represented by BDDs, every BDD node is replaced with a 2-to-1 multiplexer (MUX). The Boolean function representing a MUX is equal to $f = x \cdot \bar{s} + y \cdot s$, where s, x, and y are the input variables. The first step for the design of logic-in-memory computing circuits is to execute its logic primitive, that is, node realization, with RRAM devices. Then, a design methodology is required to implement the entire graph.

3.3.2.1 Realization of Multiplexer Using RRAM Devices

Figure 3.9 shows the IMP-based realization for MUX proposed in [5]. The realization requires six IMP operations and five RRAM devices of which three, named S, X, and Y, store the inputs and the two others, A and B, are required for operations. The corresponding implication steps of the MUX realization shown in Fig. 3.9 are as follows:

$$\mathbf{1}: S = s, X = x, Y = y, A = 0, B = 0$$
$$\mathbf{2}: a \leftarrow s \text{ IMP } a = \bar{s}$$
$$\mathbf{3}: a \leftarrow y \text{ IMP } a = \bar{y} + \bar{s}$$
$$\mathbf{4}: b \leftarrow a \text{ IMP } b = y \cdot s$$
$$\mathbf{5}: s \leftarrow x \text{ IMP } s = \bar{x} + s$$
$$\mathbf{6}: b \leftarrow s \text{ IMP } b = x \cdot \bar{s} + y \cdot s$$

In the first step, devices keeping the input variables and the two extra work switches are initialized. The remaining steps are performed by sequential IMP operations and the MUX function is realized in the sixth step.

To find the RM3-based realization of MUX, we first express the Boolean function of a multiplexer with majority gates and then simply convert it to RM3 by adding a complement attribute to each gate. For this purpose, the AND and OR operations are represented by majority gates using a constant as the third input variable, that is, 0 for AND and 1 for OR [1]. Accordingly, a MUX with input variables x, y and a

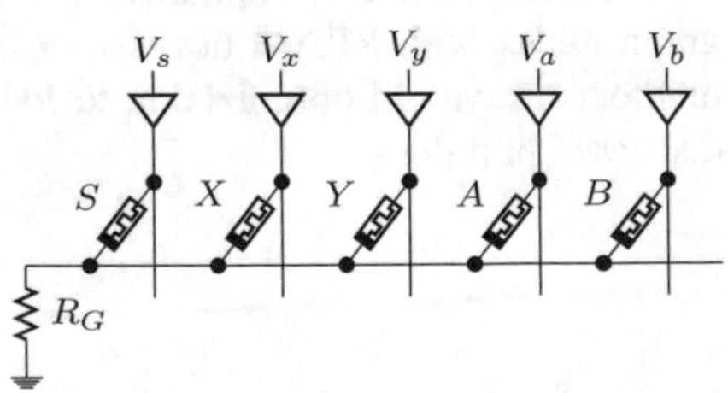

Fig. 3.9 The realization of an IMP-based MUX using RRAM devices [5]

select input s can be expressed as:

$$\begin{aligned} x \cdot \bar{s} + y \cdot s &= M(M(x, \bar{s}, 0), M(y, s, 0), 1) \\ &= M(M(x, \bar{s}, 0), M(y, 0, s), 1) \\ &= \mathrm{RM3}(\mathrm{RM3}(x, s, 0), \overline{\mathrm{RM3}}(y, 1, s), 1). \end{aligned}$$

The equations above can be executed by three RM3 operations as well as a negation. Therefore, the RM3-based realization of the MUX can be obtained by the following operations after a data loading step:

$$\begin{aligned} &\mathbf{1}: S = s, X = x, Y = y, A = 0, B = 0, C = 1 \\ &\mathbf{2}: P_A = x, Q_A = s, R_A = 0 \Rightarrow R'_A = x \cdot \bar{s} \\ &\mathbf{3}: P_S = y, Q_S = 1, R_S = s \Rightarrow R'_S = y \cdot s \\ &\mathbf{4}: P_B = 1, Q_B = s, R_B = 0 \Rightarrow R'_B = \overline{y \cdot s} \\ &\mathbf{5}: P_C = a, Q_C = b, R_C = 1 \Rightarrow R'_C = x \cdot \bar{s} + y \cdot s \end{aligned}$$

Considering area and delay, two equally important cost metrics, using IMP or RM3 does not make a difference in the circuits synthesized by the BDD-based approach as the number of RRAM devices in one realization is equal to the number of steps for the other and vice versa. Indeed, the RM3-based realization of BDD nodes allows faster circuits, while the IMP-based realization leads to circuits with smaller area consumption. Such property in both realizations can be exploited when higher efficiency in delay or area is intended.

3.3.2.2 Design Methodology for BDD-Based Synthesis

In order to escape heavy delay penalties, we assume parallelization per level for BDD-based synthesis [5, 14]. In the parallel implementation, each time one BDD level is evaluated entirely starting from the level designating the last ordered variable to the first ordered variable, the so-called root node. This is performed through transferring the computation results between successive levels, that is, using the outputs of each computed level as the inputs of the next level. Using IMP, the results of previous levels are copied wherever required within the first loading step of the next level, while for executing RM3, the results are read and then applied as voltages to the rows and columns, that is, the top and bottom electrodes of the devices.

According to the MUX realizations explained above, the number of RRAM devices required for computing by this approach is equal to five or six times the maximum number of nodes in any BDD level, when IMP-based or RM3-based realizations are used, respectively. In a similar way, the number of computational steps is six or five times the number of BDD levels. A multiple row crossbar architecture entirely based on resistive switches such as one shown in Fig. 3.10 can be used to realize the presented parallel evaluation.

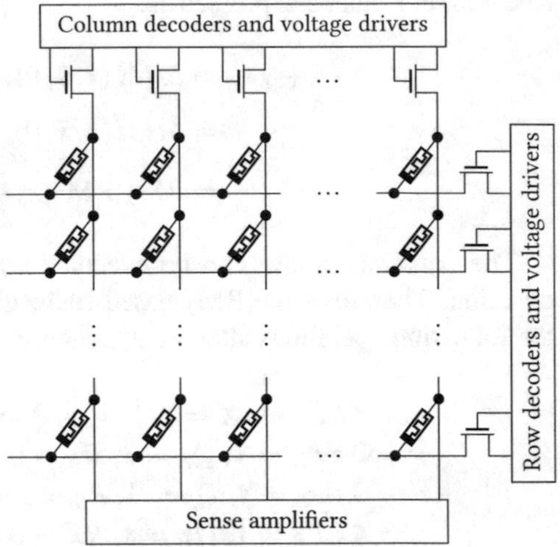

Fig. 3.10 Standard RRAM array

Although the larger part of the cost's representing area and delay of the resulting circuits are due to the number of levels and the nodes within them, some additional RRAM devices are still required. Every complemented edge in the BDD requires a NOT gate to invert its logic value. As shown in the computational steps for both IMP-based and RM3-based realizations, inverting a variable can be executed after an operation with a zero-loaded RRAM device (see step 2 in the IMP-based MUX and step 4 in the RM3-based MUX descriptions). Accordingly, for each MUX with a complemented input, an extra RRAM device should be considered and set to FALSE ($Z = 0$) that can be performed in parallel with the first loading step without any increase in the number of steps. Then, an IMP or RM3 operation should be executed to complete the logic NOT operation. All of the complemented edges in a level can be computed simultaneously that means for any level with ingoing complemented edges, only one extra step is required.

In the parallel evaluation approach, the RRAM devices keeping the outputs of each BDD level can be reused and assigned to the inputs of the next successive level. Nevertheless, the results of nodes targeting levels which are not right after their origin level might be lost during computations if their corresponding RRAM devices are rewritten by the next operations. Thus, extra RRAM devices should be allocated for such nonconsecutive fanouts to retain the result of their origin nodes to be used as an input signal of their target nodes. This will not increase the number of steps because copying the results of nodes with nonconsecutive fanouts in additional RRAM devices and using the stored value in the fanouts' targets can be performed simultaneously in the first data loading step of nodes on both sides of the fanouts.

Optimization of BDDs can be carried out as a bi-objective problem aiming at minimizing the number of RRAM devices and computational steps simultaneously, that is, finding a trade-off between the number of RRAM devices and operations of the resulting circuits [13, 14]. Figure 3.11 shows an example with two BDDs both representing a four-variable two-output Boolean function. The figure denotes

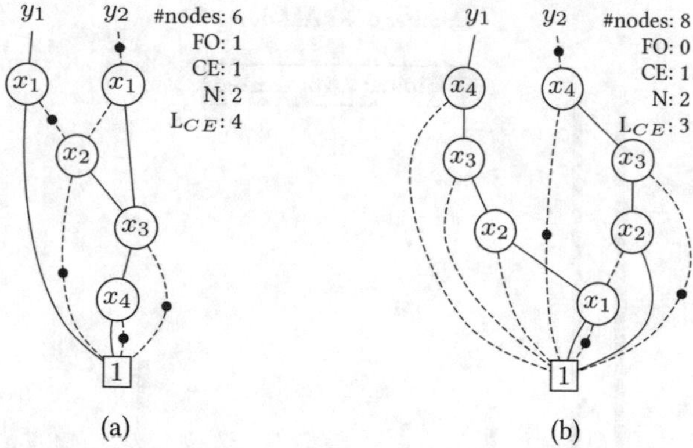

Fig. 3.11 Cost metrics of RRAM-based in-memory computing for an arbitrary BDD, (**a**) before (initial: $\#R = 12$, $\#OP = 28$) and (**b**) after optimization (optimized: $\#R = 11$, $\#OP = 27$)

the factors determining the cost metrics by FO, CE, N, and L_{CE}, which refer to the number of nonconsecutive fanouts, the maximum number of ingoing complemented edges among all levels, the maximum level size, and the number of levels with complemented edges, respectively.

As Fig. 3.11 shows, the BDD on the left has the initial ordering, whereas the second BDD has been optimized and therefore has an optimal variable ordering. The number of required RRAM devices for computing BDD levels, that is, the sum of level sizes and their ingoing complemented edges (N + CE), is equal before and after optimization since both BDDs have a maximum number of two nodes and one ingoing complemented edge per level. However, there is a nonconsecutive fanout of node x_3 targeting x_1 before optimization which requires an extra RRAM device to maintain the intermediate result. In the optimized BDD, the inputs of all of the nodes come from the consecutive levels or the constant 1 which has reduced the number of required RRAM devices by 1. The number of operations has been also reduced after optimization since one level has been released from complemented edges.

As can be seen, the numbers of RRAM devices and operations decrease although the number of BDD nodes increases. The effect of BDD optimization sounds to be too small for the example function by reducing each one of the cost metrics only by one. Nevertheless, this reduction can be much more visible for larger functions due to the higher possibility of finding BDDs with smaller number of nonconsecutive fanouts, complemented edges, and level sizes caused by larger search space.

Figure 3.12 shows the evaluated cost metrics, that is, number of RRAM devices and computational steps, for the presented BDD-based approach for logic-in-memory computing. The evaluations are performed on a set of benchmark functions selected from LGsynth91 [23]. The chart showing the number of required RRAM devices reveals that optimization has resulted in considerable improvements for most of the benchmark functions. The number of computational steps has been

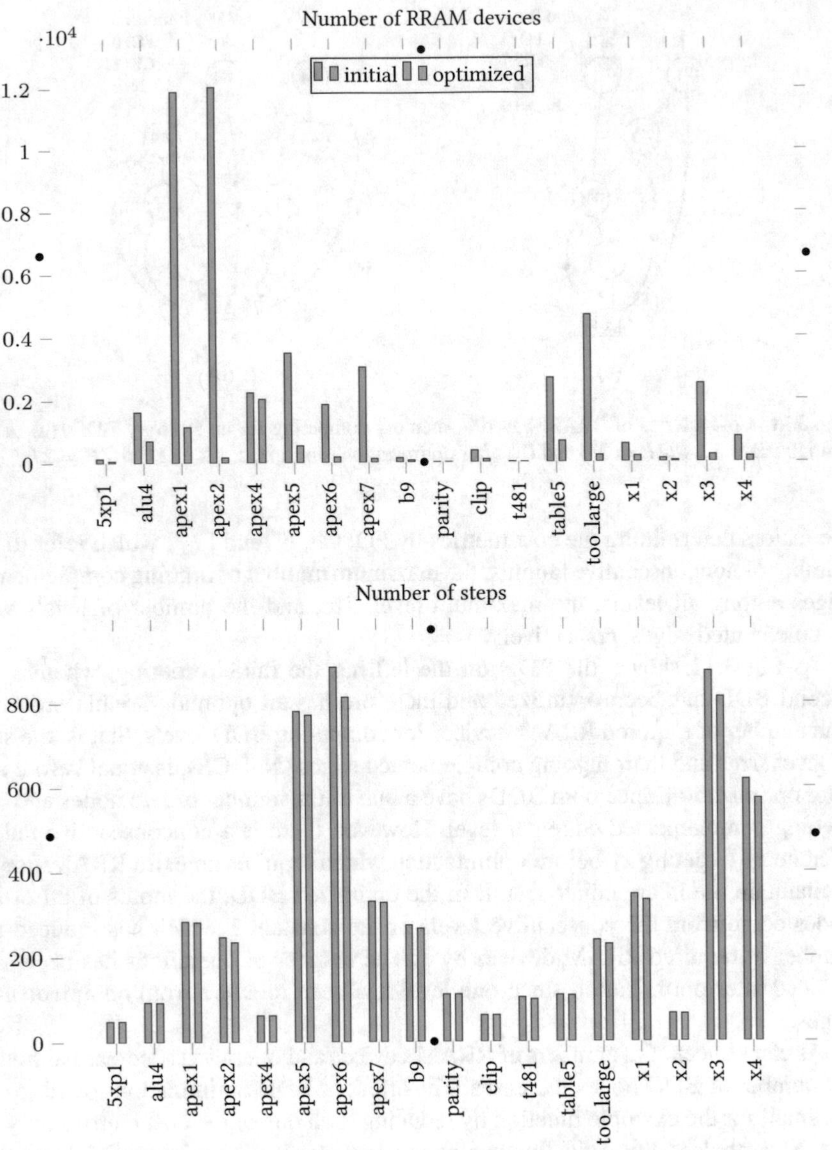

Fig. 3.12 The number of RRAM devices and computational steps required for BDD-based logic-in-memory computing evaluated before and after optimization

also decreased, however, not as much as the other cost metric. Indeed, optimization cannot noticeably lower the number of operations. As discussed before, the main contribution to the step count is the number of BDD levels, that is, the number of input variables, and hence is not changeable.

3.3.3 *AIG-Based Synthesis for In-Memory Computing*

3.3.3.1 Realization of NAND/AND Gate Using RRAM Devices

Realization of NAND gate, that is, $\overline{x \cdot y}$, using RRAM devices and based on material implication has been proposed in [4]. The proposed NAND gate in [4] corresponds to an AIG node with complemented edge and therefore can be utilized as the IMP-based realization required for AIG-based logic-in-memory computing. In this case, a negation is required for any regular, that is, non-complemented, edge in the graph. The implementation proposed in [4] requires three RRAM devices connected by a common horizontal nanowire to a load resistor, that is, structurally similar to the circuit shown in Fig 3.9 with a smaller number of devices. The three computational steps executing the NAND operation are listed below.

$$\begin{aligned}
&\mathbf{1}: X = x, Y = y, A = 0 \\
&\mathbf{2}: a \leftarrow x \text{ IMP } a = \bar{x} \\
&\mathbf{3}: a \leftarrow y \text{ IMP } a = \bar{x} + \bar{y}
\end{aligned}$$

Using RM_3, AIG can also be implemented with equal number of RRAM devices and operations. A majority operation of two variables x and y together with a constant logic value of 0 ($M(x, 0, y)$) [1] executes the AND operation. This corresponds to $RM_3(x, 1, y)$, as the second operand should be inverted. The RM_3 operation first needs to load operand y in an RRAM device, and then the AND function can be computed in the next step. The required steps are as follows:

$$\begin{aligned}
&\mathbf{1}: X = x, Y = y, A = 0 \\
&\mathbf{2}: P_A = y, Q_A = 1, R_A = 0 \Rightarrow R'_A = y \\
&\mathbf{3}: P_A = x, Q_A = 1, R_A = y \Rightarrow R'_A = x \cdot y.
\end{aligned}$$

3.3.3.2 Design Methodology for AIG-Based Synthesis

Using the discussed parallel evaluation method, an AIG is computed in a level-by-level approach such that the employed RRAM devices to evaluate the level can be reused for the next levels. Starting from the inputs of the graph, the RRAM devices in a level are released when all the required operations are executed. Then, the RRAM devices are reused for the upper level, and this procedure is continued until the target function is evaluated. Depending on the use of IMP or RM_3 in the realization, such an implementation requires as many NAND or AND gates as the maximum number of nodes in any level of the AIG. Accordingly, the corresponding number of RRAM devices is three times the number of nodes in the level with the largest size, and the number of computational steps is three times the number of levels. However, still some additional RRAM devices should be allocated for inverting the complemented or the regular edges when RM_3-based or IMP-based realizations are used, respectively.

AIGs can be optimized before being implemented on the memory array. To address the area of the resulting circuits using RRAM devices, the AIG should be optimized with respect to the number of nodes. Also, the latency of the resulting logic-in-memory computing deigns can be reduced by finding an equivalent AIG with a smaller depth, that is, the number of levels. These two cost metrics in an AIG, that is, the number of nodes and levels, conflict each other. For example, area minimization leads to worsening the latency, and, on the other hand, depth minimization increases the number of nodes in the graph. Thus, according to the application, one can choose optimization with respect to the area or delay of the resulting circuits. Nevertheless, applying optimization regarding any of the aforementioned objectives can noticeably reduce the cost metrics of RRAM-based in-memory computing [17].

Figure 3.13 shows the number of RRAM devices and computational steps required for the presented AIG-based approach for logic-in-memory computing. The results show values for both area and depth optimization methods using the IMP-based realization. As the charts in Fig. 3.13 show, the number of RRAM devices obtained by the area optimization method is mostly smaller than corresponding values resulted by depth optimization which was expected. A similar conclusion can be made when comparing the results with respect to the number of computational steps, which is smaller for most of the benchmark functions in case of the depth optimization.

In order to show the complexity of the problem, the AIG representation of one of the smallest benchmark functions used for experiments is shown in Fig. 3.14. The figure shows an AIG representing the function x_2 with ten input variables and seven output functions. Although the represented function is relatively small, its AIG representation possesses 62 nodes which means requiring 62 AND or NAND gates besides inverters for the in-memory implementation.

3.3.4 *MIG-Based Synthesis for In-Memory Computing*

3.3.4.1 Realization of Majority Gate Using RRAM Devices

Two realizations for the logic primitive of MIG, that is, the majority gate $x \cdot y + y \cdot z + x \cdot z$, are presented below based on IMP and RM_3 [15]. The following IMP-based realization of majority gate is similar to the circuit shown in Fig. 3.9 with six RRAM devices. It also requires ten sequential steps to execute the majority function.

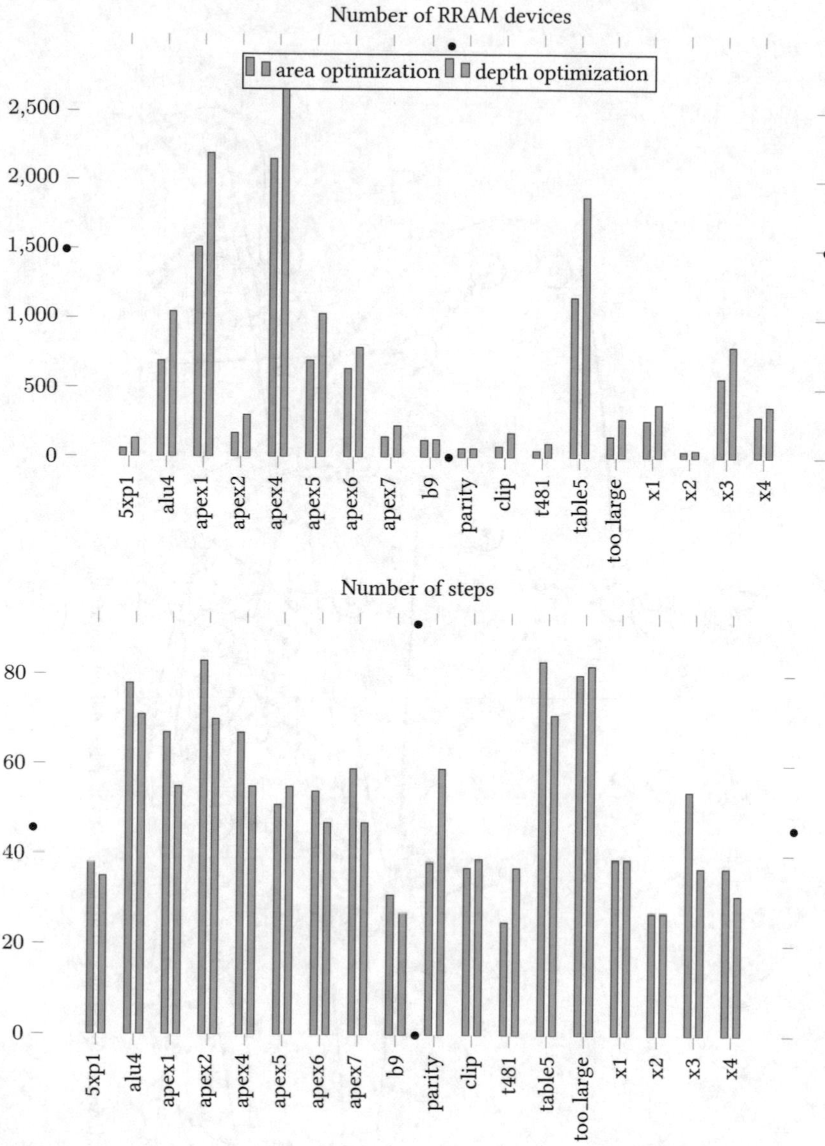

Fig. 3.13 The number of RRAM devices and computational steps required for AIG-based logic-in-memory computing evaluated for area and depth AIG optimization methods

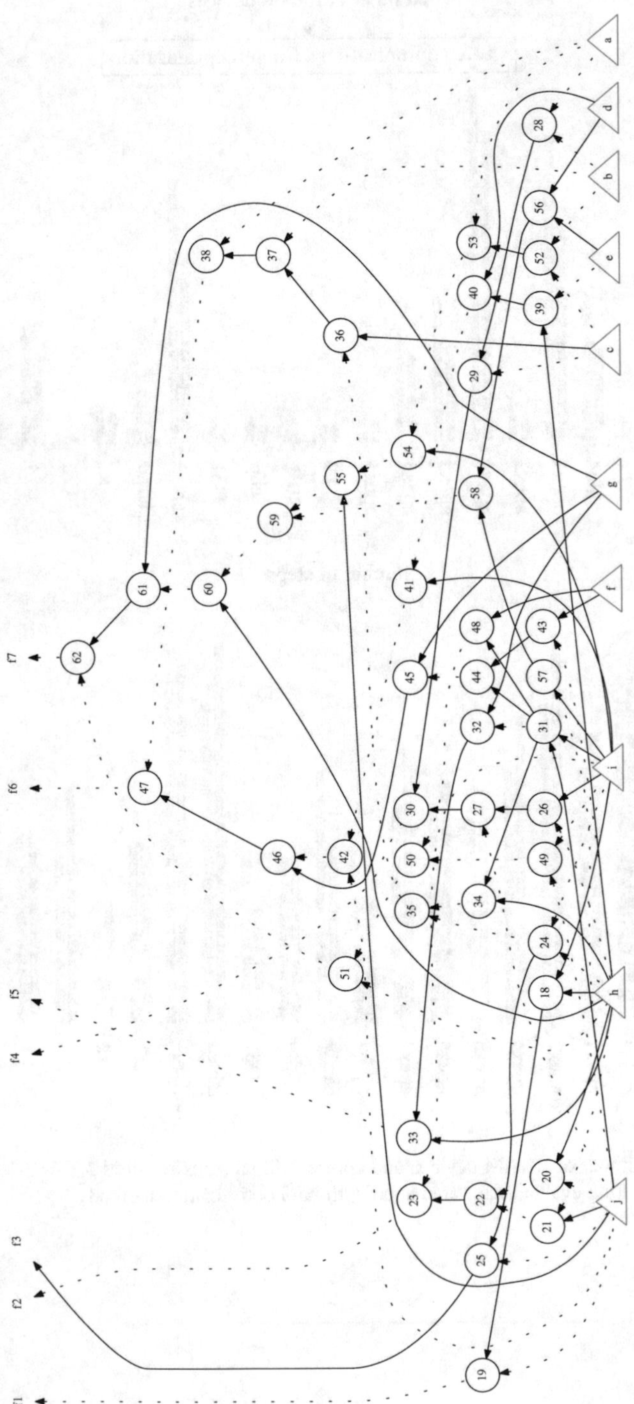

Fig. 3.14 Example AIG representation of function x_2 with ten input variables and seven outputs

The corresponding steps for executing the majority function are as follows:

$\mathbf{01}: X = x, Y = y, Z = z$ $A = 0, B = 0, C = 0$	$\mathbf{06}: c \leftarrow y \text{ IMP } c = \overline{x + y}$
$\mathbf{02}: a \leftarrow x \text{ IMP } a = \bar{x}$	$\mathbf{07}: c \leftarrow z \text{ IMP } c = \overline{x \cdot z + y \cdot z}$
$\mathbf{03}: b \leftarrow y \text{ IMP } b = \bar{y}$	$\mathbf{08}: a = 0$
$\mathbf{04}: y \leftarrow a \text{ IMP } y = x + y$	$\mathbf{09}: a \leftarrow b \text{ IMP } a = x \cdot y$
$\mathbf{05}: b \leftarrow x \text{ IMP } b = \bar{x} + \bar{y}$	$\mathbf{10}: a \leftarrow c \text{ IMP } a = x \cdot y + y \cdot z + x \cdot z.$

Three RRAM devices denoted by X, Y, and Z keep input variables, and the remaining three other RRAM devices A, B, and C are required for retaining the intermediate results and the final output. In the first step, the input variables are loaded, and the other RRAM devices are assigned to FALSE to be used later for the next operations. Another FALSE operation is also performed in step 8 to clear an RRAM device which is not required anymore for inverting an intermediate result which is not required anymore. Finally, the Boolean function representing a majority gate is executed by implying results from the seventh and ninth step.

It is obvious that the RM3-based majority gate can be realized with a smaller number of RRAM devices and computational steps due to benefiting from the discussed built-in majority property. Using RM3, the majority gate will require only four RRAM devices, and the majority function can be executed within only three steps. The computational steps for the RM3-based realization are as follows:

$$\mathbf{1}: X = x, Y = y, Z = z, A = 0$$
$$\mathbf{2}: P_A = 1, Q_A = y, R_A = 0 \Rightarrow R'_A = \bar{y}$$
$$\mathbf{3}: P_Z = x, Q_Z = \bar{y}, R_Z = z \Rightarrow R'_Z = M(x, y, z).$$

In the first step, the initial values of input variables as well as an additional RRAM device are loaded. Step 2 executes the required NOT operation in RRAM device A, and the last step computes the majority function by use of RM3 at RRAM device Z.

3.3.4.2 Design Methodology for MIG-Based Synthesis

The number of RRAM devices and operations for the proposed MIG-based synthesis approach can be calculated similarly to that performed for AIGs with RM3-based realization but with respect to the realizations given for MIGs' logic primitive. Since both IMP-based and RM3-based realizations proposed for MIGs represent majority gate without an extra negation, the same formula can be used for them with different constant factors addressing the number of RRAM devices and operations required by each realization [15]. This means that to compute every MIG level, at least a number of ten or three steps are required, when IMP-based or RM3-based realizations are used, respectively. The number of RRAM devices also is a

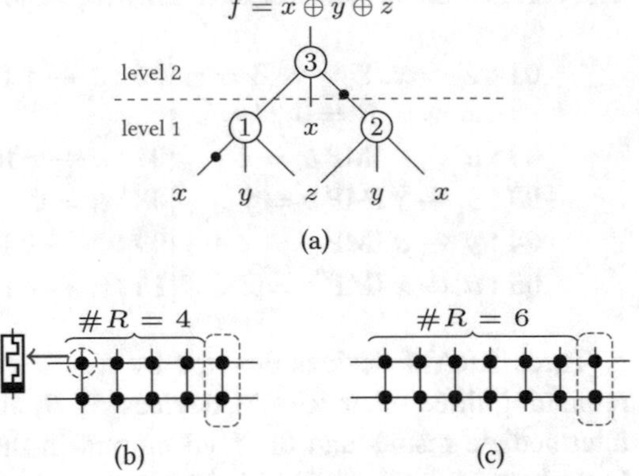

Fig. 3.15 (**a**) Example MIG for logic-in-memory implementation. (**b**) Upper-bound RRAM array for implementation using RM3-based realization. (**c**) Upper-bound RRAM array for implementation using IMP-based realization

multiplication of the number of nodes in the levels to the RRAM devices required for each node according to the employed realization.

MIG optimization can be performed by applying a set of valid transformations, for example, in [1], to an existing MIG to find an equivalent MIG that is more efficient with respect to the number of RRAM devices and steps. MIG optimization algorithms have been presented for tackling the cost metrics of logic synthesis with RRAM devices. In this section, we avoid the technical details of the MIG optimization algorithms and refer the reader to [17] for an overall view. It is worth mentioning that MIG optimization can consider both cost metrics simultaneously [15], while it can only aim at reducing the number of operations [15] or RRAM devices [18].

For a better understanding of the explanations above, the implementation steps of an example MIG are shown in this section, and its requirements including the number of RRAM devices, computational steps, and dimensions of the memory array are discussed in detail.

Figure 3.15a shows an MIG which represents a three-input XOR gate, that is, $\overline{xy}z + \bar{x}y\bar{z} + x\overline{yz} + xyz$. Both RM3-based and IMP-based implementations for in-memory computing logic circuits can be executed on a standard crossbar architecture as shown in Fig. 3.10. However, much smaller dimensions are required in this case which are illustrated for both node realizations in Fig. 3.15b, c. For implementation based on the presented parallel methodology, an entire row should be allocated for computing a single graph node which designates the corresponding logic primitive, for example, a majority gate if MIG is used for synthesis. Accordingly, to compute an entire level of a logic representation, all nodes of the level should be computed in parallel within separate rows of the memory array. This means that a number of rows equal to the maximum level size in the entire graph is required. For example, for a logic representation whose largest level has four nodes, a memory array with at least four rows is needed, independent of the type of the utilized representation or the basic operation of RM3 or IMP. Nevertheless, the number of RRAM devices at each row, that is, the number of columns, is determined by the RRAM-based node realization of the exploited logic representation.

RM3-Based Implementation According to the RM3-based realization of majority gate and the discussions before, the computation of the MIG shown in Fig. 3.15 can be performed using a maximum of nine RRAM devices, since the critical level needs 2×4 for its nodes and one more RRAM for the ingoing complemented edge. Also, each level can require three operations (2×3), which results in a total of eight operations for the whole MIG considering the presence of complemented edges at both levels. However, the RM3-based implementation can actually be executed much more efficiently with respect to both time and area.

The upper-bound memory array required for implementation is shown in Fig. 3.15b which has two rows as the target MIG has a maximum level size of two. Each row consists of four RRAM devices to compute a node and one additional device to be used in case of having a complemented edge. This way, a maximum of two ingoing edges for an MIG node can be complemented after rewriting, from which one can directly be used as the second inverted operand of RM3 and, thus, only one needs to be negated. The RRAM devices allocated for the complemented edges are displayed in red-dashed surrounds at the end of the rows. The implementation steps for the RM3-based computation of the MIG shown in Fig. 3.15 are listed below:

Initialization : $R_{ij} = 0 : Q_{ij} = 1,\ P_{ij} = 0;$

1 : Loading third operands $Q_1 = Q_2 = 0,\ P_1 = P_2 = z;$
$R_{11} : \text{RM3}(z, 0, 0) = M(z, 1, 0) = z;$
$R_{21} : \text{RM3}(z, 0, 0) = M(z, 1, 0) = z;$

2 : Negation for node 2 $Q_1 = Q_2 = x,\ P_1 = x,\ P_2 = 1;$
$R_{25} : \text{RM3}(1, x, 0) = M(1, x, 0) = x;$

3 : Computing level 1 <u>node 1:</u>$P_1 = y,\ Q_1 = x,\ R_{11} = z$
$R_{11} : \text{RM3}(y, x, z) = M(y, \bar{x}, z);$
<u>node 2:</u>$P_1 = y,\ Q_2 = \bar{x}(@R_{25}),\ R_{21} = z;$
$R_{21} : \text{RM3}(y, \bar{x}, z) = M(y, x, z);$

4 : Computing level 2 (f) $P_1 = x,\ Q_1 = @R_{21},\ R_{11} = M(\bar{x}, y, z);$
$R_{11} : \text{RM3}(x, @R_{21}, @R_{11}) = M(x, \overline{@R_{21}}, @R_{11}) :$
$M(M(\bar{x}, y, z), x, \overline{M}(x, y, z));$

It is assumed that all of the RRAM devices are first loaded with zero for initialization. Then, step 1 starts to compute the nodes 1 and 2 in level 1 (see Fig. 3.15) by loading the variable z. As said before, every node of the level should be computed in a separate crossbar row. Accordingly, nodes 1 and 2 are, respectively, computed in rows 1 and 2 by selecting R_{11} and R_{21}, where i and j in R_{ij} refer to row and column, respectively. For this purpose, the corresponding third operands, that is, the destinations of the operations, are loaded first. Then, the primary inputs are read from memory and applied to the corresponding row and columns to execute the operations.

As the implementation steps show, the root function f is computed using only three RRAM devices from the entire allocated array in Fig. 3.15b. This reduction in the number of RRAM devices compared to the expected value has two main reasons. One reason is that the RM_3-based realization of majority gate also allocates RRAM devices for the first and the second operands which are not required always, while IMP needs all variables to be accessible on the same row. Therefore, the number of RRAM devices provided by the array shown in Fig. 3.15b is more than what is actually needed. Furthermore, in the RM_3-based realization of majority gate, it was simply assumed that the second operand needs to be inverted for which an RRAM device was considered which can be avoided in case that the MIG node already has a single ingoing complemented edge. For example, in the MIG shown in Fig. 3.15, nodes 1 and 3 (the root node) are ideal for RM_3 due to possessing a single complemented edge, but node 2 requires one negation which needs to be performed first.

IMP-Based Implementation The required array for the IMP-based implementation is shown in Fig. 3.15c which has one extra RRAM at the end of each row for the complemented edges. According to the IMP-based realization of majority gate, the example MIG shown in Fig. 3.15a with a maximum level size of 2 needs an upper bound of 12 (2×6) RRAM devices placed in two rows in addition to one more for the ingoing complemented edge. As the parallel evaluation methodology suggests, the computation needs 22 steps, 2×10 for the two levels plus two more steps for the complemented edges, including the IMP operations and the loads.

The required steps for the IMP-based implementation of the MIG shown in Fig. 3.15 are listed in the following:

Initialization : $R_{ij} = 0;$

1 : Loading variables for level 1: $R_{11} = x,\ R_{12} = y,\ R_{13} = z;$
$R_{21} = x,\ R_{22} = y,\ R_{23} = z;$

2 : Negation for node 1: $R_{17} \leftarrow x\mathrm{IMP}R_{17} : R_{17} = \bar{x};$

3–11 : Computinglevel1 : $\underline{\text{node1}: }R_{14} = M(\bar{x}, y, z);$
$\underline{\text{node2}: }R_{24} : M(x, y, z);$

12 : Loading variables for level 2: $R_{11} = x,\ R_{12} = M(x, y, z),\ R_{13} = M(\bar{x}, y, z)$
$R_{14} = R_{15} = R_{16} = R_{17} = 0;$

13 : Negation for node 3: $R_{17} \leftarrow R_{12}\mathrm{IMP}R_{17} :$
$R_{17} = \overline{R_{12}} = \overline{M}(x, y, z);$

14–22 : Computing level 2 (f) : $R_{14} = M(M(\bar{x}, y, z), x, \overline{M}(x, y, z));$

Dissimilar to RM_3, IMP needs all variables used for computation to be stored in the same horizontal line. This means that there may be a need to have several copies of primary inputs or intermediate results at different rows simultaneously, as shown in step 1, where the variables of nodes 1 and 2 are loaded into RRAM devices in both rows. Step 2 computes the complemented edge of node 1 in the seventh RRAM device considered for this case at the end of the first row. Steps 3–11 compute both nodes at level 1 and store the results in the fourth RRAM device at the corresponding crossbar row. The same procedure continues to compute the second level, which only consists of the MIG root node. Two out of the three inputs of node 3 are intermediate results, which have to be first read and then copied into the corresponding RRAM devices at row 1 besides other input and work devices as shown in step 12. In step 13, the complemented edge originating at node 2 is negated, and then root node is computed in step 22.

Figure 3.16 shows results of the MIG-based synthesis using the RM_3-based realization. The charts show the number of RRAM devices and computational steps obtained by three different MIG optimization algorithms [17], that is, bi-objective algorithm aiming to find a trade-off between both cost metric, and two single-objective algorithms targeting either the number of RRAM devices or the number of steps. As the figure shows, the bar showing the values for the bi-objective optimization mostly falls between the minimum and maximum values obtained by the other algorithms. For example, the number of RRAM devices obtained by the bi-objective algorithm is smaller than obtained by the step-minimization algorithm but higher than the same value resulted after RRAM minimization.

3.4 Logic-In-Memory Computer Architecture

As shown in Sect. 3.3, logical functions can be performed within RRAM which unifies the memory and processing and eliminates the need for data transfer in the von Neumann architecture. A *Programmable Logic-in-Memory* (PLiM) computer architecture was proposed in [7]. The PLiM architecture consists of a regular resistive memory, which is also capable of performing logic operations, and a simple control unit to execute the operations. The memory includes several banks of RRAM arrays (see Fig. 3.10) which are connected to the PLiM controller block as shown in Fig. 3.17.

The PLiM controller is a synchronous block, that is, works based on a centralized clock signal, that allows the RRAM arrays to function in computation or memory modes. The transition between the computation mode and the memory mode is controlled by the *Logic-in-Memory* (LiM) input. When the LiM control signal is equal to one, PLiM works in the computation mode which is managed by the controller. The computation mode runs a sequential execution of a given set of instructions, that is, RM_3 operations, that represent a program. The program is stored on the memory array, which is automatically updated when the program is executed. When the LiM is equal to zero, the controller is off and the architecture

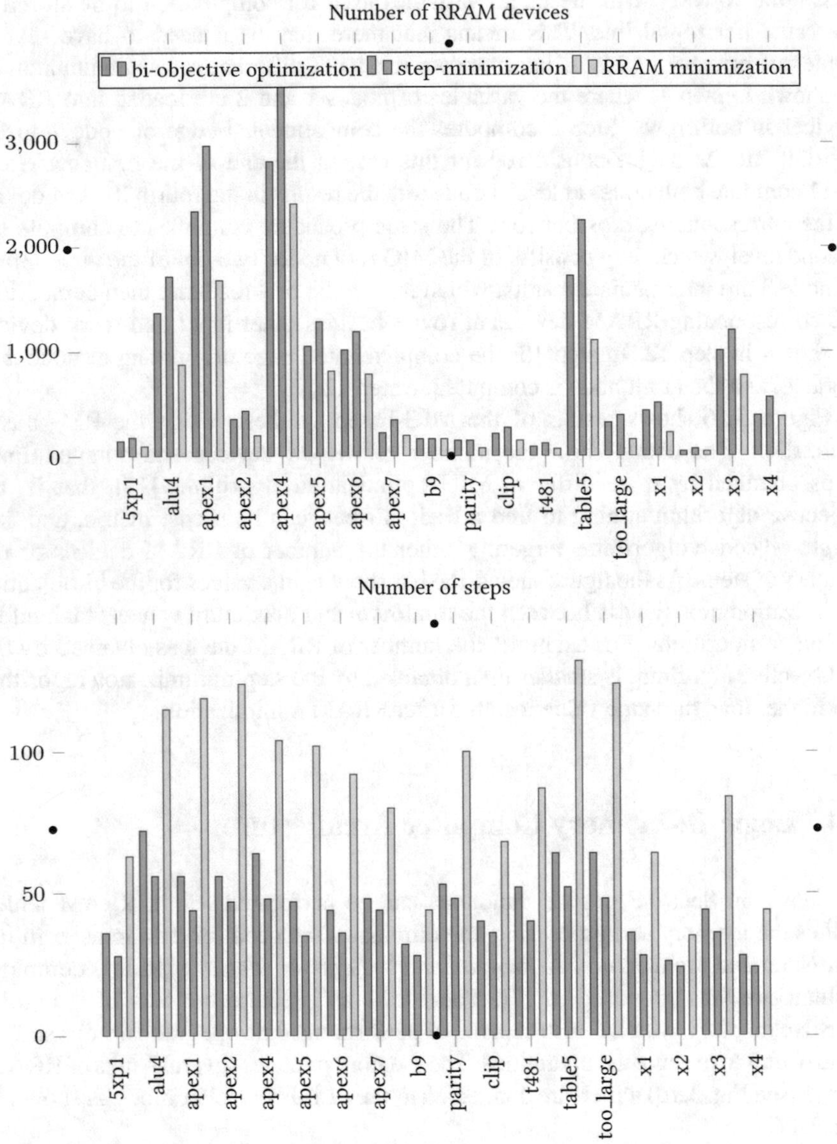

Fig. 3.16 The number of RRAM devices and computational steps required for MIG-based logic-in-memory computing evaluated for three different optimization schemes

works as pure memory. The PLiM architecture deals with a single instruction at each clock cycle which simplifies the controller [7].

Every single PLiM instruction is an $\mathrm{RM}_3(P, Q, R)$ operation, while a PLiM program is referred to a sequence of instructions. Performing the instruction simply means loading the bit-level values of P and Q from memory and applying them

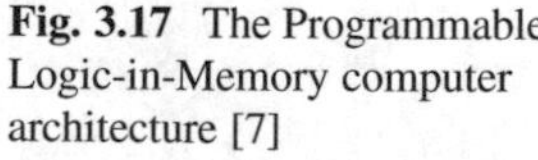

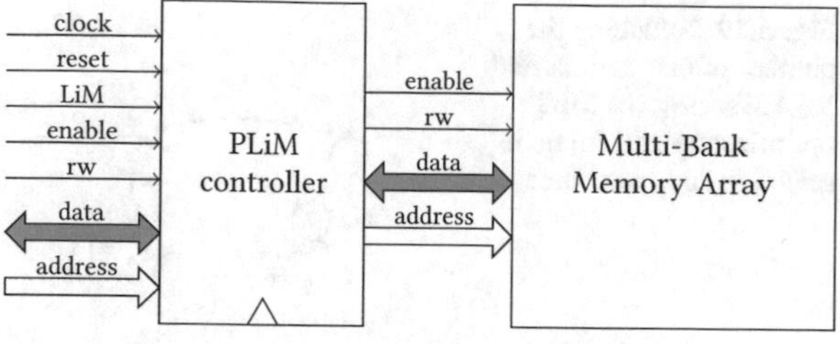

Fig. 3.17 The Programmable Logic-in-Memory computer architecture [7]

ZERO(z)	$z \leftarrow \mathrm{RM}_3(0, 1, z) = \langle 00z \rangle = 0$
ONE(z)	$z \leftarrow \mathrm{RM}_3(1, 0, z) = \langle 11z \rangle = 1$
BUF(a, z)	ZERO(z); $z \leftarrow \mathrm{RM}_3(a, 0, z) = \langle a10 \rangle = a$
NOT(a, z)	ZERO(z); $z \leftarrow \mathrm{RM}_3(1, a, z) = \langle 1\bar{a}0 \rangle = \bar{a}$
RM(a, b, z)	$z \leftarrow \mathrm{RM}_3(a, b, z)$

Fig. 3.18 The PLiM commands [19]

to the RRAM device *R*. The instruction itself is stored on the same memory bank which should be loaded first. Then, the operands are loaded from the memory, and finally, the operands are applied to the electrodes of the destination RRAM, i.e. the row and column in which the device is located. When the operation is performed, the resistance state of the destination RRAM is updated by the result of the instruction.

Any Boolean function can be computed with a sequence of RM_3 instructions. Figure 3.18 shows a set of primitive logical functions sufficient to compute any function which we refer to as PLiM commands. Each command consists of one, for example, ZERO(z), or two, for example, BUF(a, z), RM_3 instructions, where z is the RRAM device storing the result of the command. Accordingly, PLiM can be programmed to compute an arbitrary function within a finite sequence of instructions. In contrast to the processors in modern computer architectures where computations are performed on embedded circuits and systems, the PLiM computes on banks of RRAM crossbar arrays (see Fig. 3.10) which makes it programmable for any purpose without any change in hardware or wiring. However, the procedure to design for PLiM can algorithmically borrow from classical hardware synthesis such as shown in Sect. 3.3.

Since the PLiM instructions are RM_3 operations, MIGs are efficiently used to represent PLiM programs [7]. Therefore, a program is implemented similarly to the MIG-based approach explained in Sect. 3.3.4 with the difference that the PLiM instructions are evaluated fully sequentially. In other words, a single MIG node is computed at each clock cycle, while the design methodology presented in Sect. 3.3.4 computes an entire level of a given MIG in parallel.

Several issues should be properly addressed during translation of an MIG into PLiM understandable commands or in other words compilation for PLiM. The order of the MIG nodes selected for execution highly matters as it affects the required number of RRAM devices and the number of instructions. Even for computing a certain node, different selections of operands can result in different number of RRAMs and steps [18, 19]. Also, allocation of RRAM devices for computations during translating a given MIG can result in an uneven distribution of writes over

Fig. 3.19 Reducing the number of instructions and RRAMs, after (**a**) MIG optimization and (**b**) node selection and translation

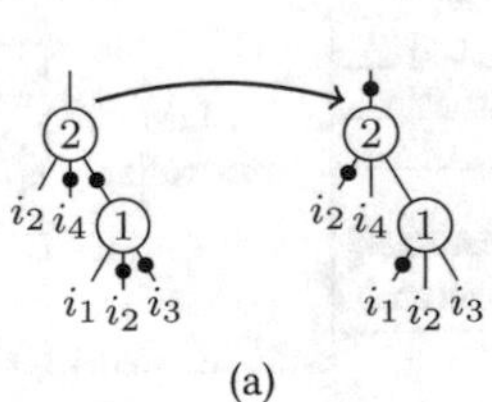

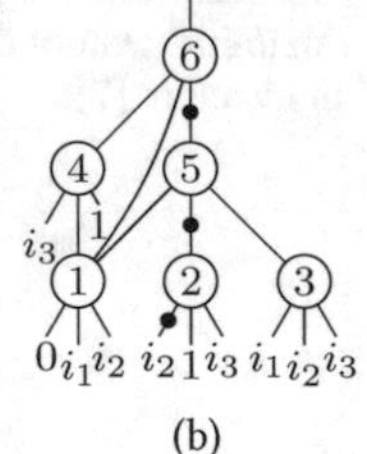

the memory array [16]. The unbalanced write traffic causes some devices to pass the limit of the number of writes earlier and cease to work reliably, that is, the so-called memory wear-out. In general, RRAM devices can endure a limited number of write cycles about 10^{10} [11] to 10^{11} [9], in the best cases. This section explains the different stages required to obtain a PLiM program from a given MIG and discusses how the aforementioned issues can be properly addressed during compilation.

Here, we show an example to explain how the characteristics of an MIG affect the resulting PLiM programs with respect to the mentioned metrics. Figure 3.19a shows a small MIG before optimization (the MIG on the left) and after optimization (the MIG on the right). As shown in the figure, optimization has changed the position and the number of inverters in the graph. The PLiM commands for both of the MIGs are listed in the following:

$\texttt{NOT}(i_3, z_1)$	$\texttt{BUF}(i_3, z_1)$
$1 \triangleright \texttt{RM}(i_1, i_2, z_1)$	$1 \triangleright \texttt{RM}(i_2, i_1, z_1)$
$\texttt{NOT}(z_1, z_2)$	$2 \triangleright \texttt{RM}(i_2, i_4, z_1)$
$2 \triangleright \texttt{RM}(i_2, i_4, z_2)$	

The commands given on the left side represent the PLiM program to compute the initial MIG, while the commands on the right represent the MIG after optimization. Before optimization, the MIG shown in Fig. 3.19 requires more computational steps and two RRAM devices, z_1 and z_2, to be implemented on PLiM. Although the optimization does not reduce the number of nodes and just affects the complemented edges, the required number of RRAM devices and commands both decrease by one.

As mentioned before, not only the MIG structure has an effect on the PLiM program, but also the order in which nodes are translated as well as the selection of the node's inputs as operands P, Q, and destination R in the RM_3 instruction. As example, consider the MIG in Fig. 3.19b. Translating it in a naïve way, i.e., in order of their node indices and selecting the RM_3 operands and destination in order of their inputs (from left to right), will result in PLiM program shown in the following on the left side which needs 13 commands and 7 RRAM devices. By changing the order in which the nodes are translated and also the order in which operands are selected for the RM_3 instructions, a shorter program requiring 11 commands and only 4 RRAM devices can be found for the same MIG representation on the right side:

$\mathrm{NOT}(i_1, z_1)$	$3 \rhd \mathrm{RM}(i_1, z_4, z_5)$	$\mathrm{BUF}(i_2, z_1)$	$3 \rhd \mathrm{RM}(i_1, z_3, z_4)$
$\mathrm{BUF}(i_2, z_2)$	$\mathrm{NOT}(i_3, z_6)$	$1 \rhd \mathrm{RM}(i_1, 1, z_1)$	$5 \rhd \mathrm{RM}(z_1, z_2, z_4)$
$1 \rhd \mathrm{RM}(0, z_1, z_2)$	$\mathrm{ONE}(z_7)$	$\mathrm{ONE}(z_2)$	$\mathrm{BUF}(i_3, z_2)$
$\mathrm{BUF}(i_3, z_3)$	$4 \rhd \mathrm{RM}(z_2, z_6, z_7)$	$2 \rhd \mathrm{RM}(i_3, i_2, z_2)$	$4 \rhd \mathrm{RM}(z_1, 0, z_2)$
$2 \rhd \mathrm{RM}(1, i_2, z_3)$	$5 \rhd \mathrm{RM}(z_2, z_3, z_5)$	$\mathrm{NOT}(i_2, z_3)$	$6 \rhd \mathrm{RM}(z_1, z_4, z_2)$
$\mathrm{NOT}(i_2, z_4)$	$6 \rhd \mathrm{RM}(z_7, z_5, z_2)$	$\mathrm{BUF}(i_3, z_4)$	
$\mathrm{BUF}(i_3, z_5)$			

As discussed before, the way an MIG is translated into a PLiM program also affects the write traffic of the allocated memory. During PLiM compilation, write endurance can be easily considered whenever an RRAM device is requested such that the device with the smallest write count is returned. This is a simple but also an essential technique to lower the deviation of writes over the memory used for computation. However, to obtain a good write balance, the MIG characteristics causing imbalance in the distribution of writes have to also be properly addressed.

For example, computing the MIG shown in Fig. 3.20 by PLiM, the computations cannot be balanced completely over all allocated RRAMs unless additional instructions and RRAM devices are consumed. We assume that node A is already computed and node B is the best candidate for PLiM to be computed next with respect to the number of instructions and required devices. In this case, the compiler chooses the RRAM storing the value of node A, let us call it X, as the destination of the RM_3 instruction. This is because the RRAMs storing the two other children nodes of B have more than one fanout. Selecting a node with multiple fanouts as the destination of RM_3 will require two extra instructions and one extra RRAM device to copy the values of the nodes to use them later at their other fanout targets. Therefore, node A is written again regardless of its current number of writes, and its content is replaced by the result of computation of node B. A similar situation occurs when computing node C. To compute node C, the compiler first sets the only complemented child shown with a dotted edge, that is, node D, as the second operand of RM_3 for the sake of efficiency. Between the two other remaining children, only node B has a single fanout whose value is stored in RRAM X. Accordingly, again X will be selected as the destination of operation in order to avoid extra costs, while the RRAM keeping the value of node D might have much smaller write count.

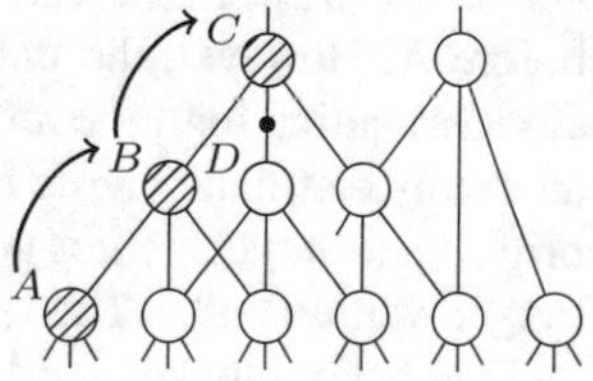

Fig. 3.20 Example MIG vulnerable to unbalanced write caused by PLiM's area latency considerations

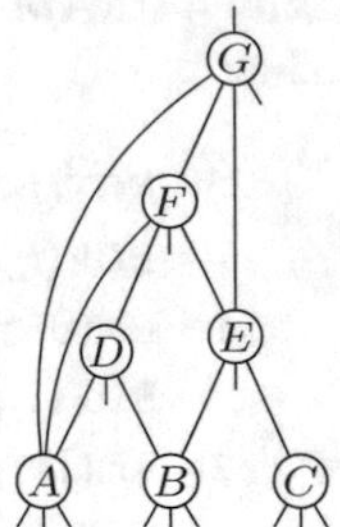

Fig. 3.21 Example MIG vulnerable to unbalanced write due to possessing nodes with long storage durations

As shown in the example above, rewriting the same RRAM repeatedly can continue for a large number of node computations depending on the situation of single fanouts and complemented edges in the MIG. This condition causes an unbalanced write distribution which cannot be controlled without extra costs in terms of execution time and area.

The structure of the MIG can help to distribute the writes more evenly if RRAMs in use are released and reused with a similar frequency. Indeed, the issue with endurance management happens when some RRAM devices are blocked for a long time and some others are rewritten very often. In such a situation, the number of writes for the RRAMs used to execute the PLiM program can vary widely. Thus, some RRAM devices reach to the end of their lifetime much faster than others which shortens the normal duration that the PLiM computer can work reliably.

Figure 3.21 shows an MIG with the problem explained above. Let us assume that node A is already computed and X_A designates the RRAM keeping its value. It is obvious that X_A is still required and cannot be released until node G is computed. Node A targets several nodes in higher levels which means longer waiting time for X_A, while all other nodes only target the nodes in the very next levels. In comparison, the RRAMs storing the values of nodes B and C, indicated by X_B and X_C, respectively, can be released when nodes D and E are computed, and one of them can be rewritten again to compute node F. X_B and X_C can also be rewritten before by the values of their target nodes D and E. Considering that the initial number of writes has been equal for all the mentioned RRAMs, X_B and X_C might have been rewritten at least one or two times more than X_A, which we call a blocked RRAM, when the root node G is computed.

The impact of blocked RRAMs on write traffic can be much more noticeable in large MIGs with high-level differences. Nevertheless, write balance can be enhanced if the nodes with long waiting time are computed as late as possible. In the example above, the write traffic can be managed if the compiler computes nodes B and C before A. However, the unbalanced write distribution caused by blocked RRAM devices waiting for quite long time cannot be eliminated but only decreased. In fact, an evenly distributed write traffic for an MIG with large differences between fanout origins and targets is not possible by only postponing computation of nodes with longer waiting times. The sequential nature of PLiM architecture anyway imposes a waiting list of blocked RRAM devices which cannot be released until the root node or nodes close to it are computed.

According to the MIG examples and the discussions above, considering the cost metrics of the PLiM programs, that is, the number of required RRAM devices and instructions, and the write traffic over the memory arrays can be conflicting objectives which can be addressed during both stages of MIG optimization and compilation [16]. Figure 3.22 shows the number of instructions and RRAM devices

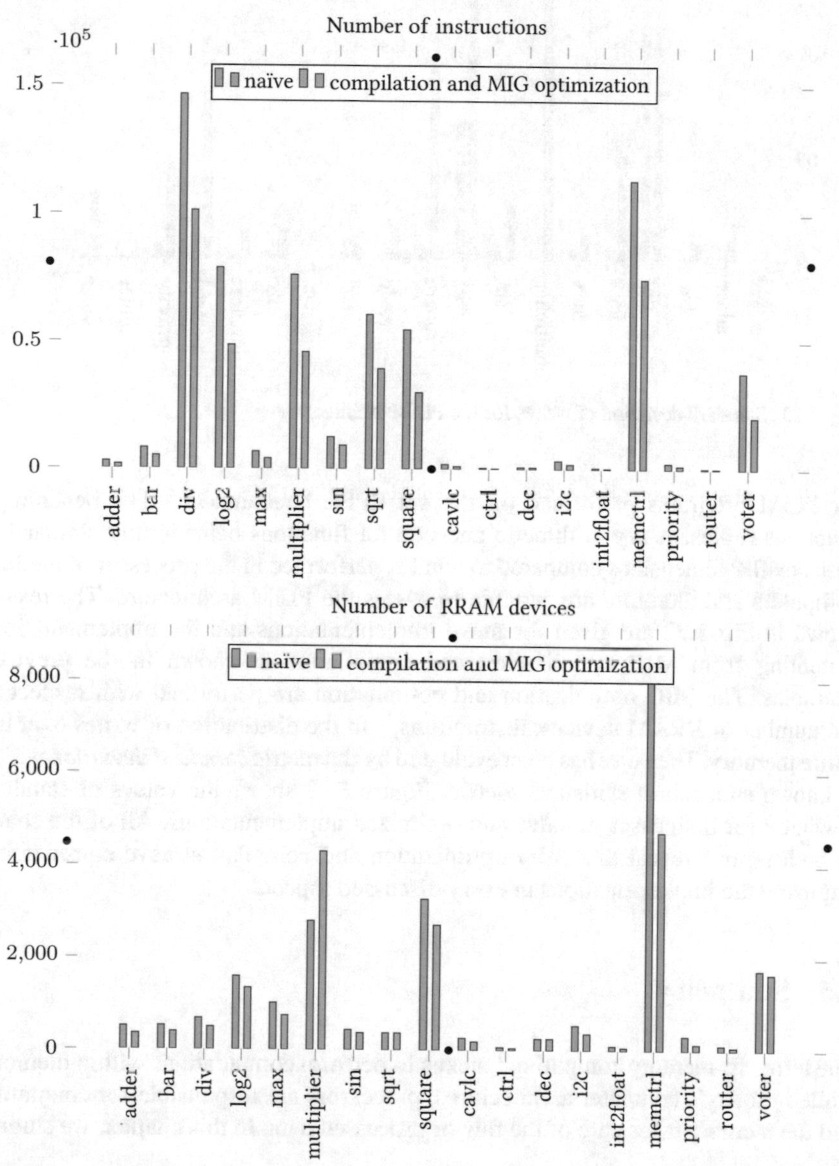

Fig. 3.22 The number of instructions and RRAM devices required by the PLiM computer architecture

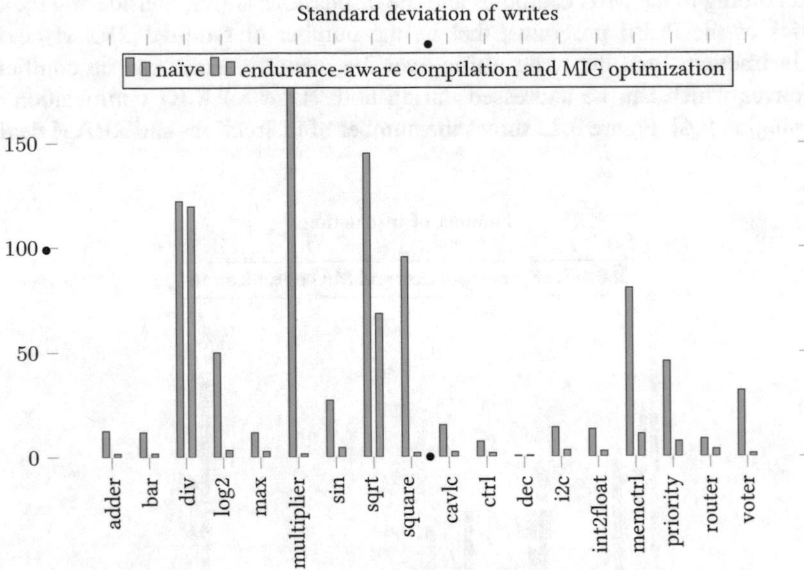

Fig. 3.23 Standard deviation of writes for the PLiM architecture

for PLiM programs performed on the set EPFL benchmarks.[4] The benchmark functions represent key arithmetic and control functions behaviorally similar but with smaller dimensions compared to what is performed in the processor of modern computers and therefore are suitable to assess the PLiM architecture. The results shown in Fig. 3.22 are given for naïve implementations and the implementations benefiting from MIG optimization and compilation as shown in the previous examples. The MIG optimization and compilation are performed with respect to the number of RRAM devices, instructions, and the distribution of writes over the entire memory. The latter has been evaluated by the metric *standard deviation* which is known as a robust statistical metric. Figure 3.23 shows the values of standard deviation for both cases of naïve and optimized implementations. All of the charts in both figures reveal that MIG optimization and compilation have considerably improved the implementations in every discussed aspect.

3.5 Summary

The term "in-memory computing" means to perform computations within memory, while in today's computer architectures, processors are responsible for computing and the memory takes care of the raw or processed data. In this chapter, we studied

[4] http://lsi.epfl.ch/benchmarks.

in-memory computing enabled by an emerging memory technology called *Resistive Random-Access Memory* (RRAM) from two perspectives. The first part of the chapter shows how some classical approaches for logic design can be exploited to compute within RRAM devices. The presented approach employs traditional and recent logic representations which are already proven to be efficient in classical logic synthesis both theoretically and experimentally. The design procedure starts with finding efficient realizations with RRAM devices for the primitive logical computation cell of each representation and then continues based on a design methodology for implementation on RRAM array. The approach also addresses optimization with respect to the area and latency.

The second part of the chapter introduces an in-memory computer architecture which is programmable and thus can compute any arbitrary logical function. In this part, we provide a step-by-step and simplified procedure to obtain efficient programs considering factors such as latency, area, and reliability for the in-memory computer architecture.

References

1. Amarù, L.G., Gaillardon, P.-E., Micheli, G.D.: Majority-inverter graph: a novel data-structure and algorithms for efficient logic optimization. In: Design Automation Conference, pp. 194:1–194:6 (2014)
2. Amarù, L.G., Gaillardon, P.-E., De Micheli, G.: Majority-inverter graph: a new paradigm for logic optimization. IEEE Trans. Comput. Aided Des. Integr. Circuits Syst. **35**, 806–819 (2015)
3. Bollig, B., Wegener, I.: Improving the variable ordering of OBDDs is NP-complete. IEEE Trans. Comput. **45**(9), 993–1002 (1996)
4. Borghetti, J., Snider, G.S., Kuekes, P.J., Yang, J.J., Stewart, D.R., Williams, R.S.: Memristive switches enable stateful logic operations via material implication. Nature **464**, 873–876 (2010)
5. Chakraborti, S., Chowdhary, P., Datta, K., Sengupta, I.: BDD based synthesis of Boolean functions using memristors. In: 2014 9th International Design and Test Symposium (IDT), pp. 136–141 (2014)
6. Drechsler, R., Sieling, D.: Binary decision diagrams in theory and practice. Int. J. Softw. Tools Technol. Trans. **3**(2), 112–136 (2001)
7. Gaillardon, P.-E., Amarù, L., Siemon, A., Linn, E., Waser, R., Chattopadhyay, A., De Micheli, G.: The programmable logic-in-memory (PLiM) computer. In: Design Automation & Test in Europe, pp. 427–432 (2016)
8. Hamdioui, S., Taouil, M., Haron, N.Z.: Testing open defects in memristor-based memories. IEEE Trans. Comput. **64**(1), 247–259 (2015)
9. Kim, Y.B., Lee, S.R., Lee, D., Lee, C.B., Chang, M., Hur, J.H., Lee, M.J., Park, G.S., Kim, C.J., Chung, U.I., Yoo, I.K., Kim, K.: Bi-layered rram with unlimited endurance and extremely uniform switching. In: Symposium on VLSI Technology, pp. 52–53 (2011)
10. Kuehlmann, A., Paruthi, V., Krohm, F., Ganai, M.K.: Robust Boolean reasoning for equivalence checking and functional property verification. IEEE Trans. Comput.-Aided Des. Integr. Circuits Syst. **21**(12), 1377–1394 (2002)
11. Lee, H.Y., Chen, Y.S., Chen, P.S., Gu, P.Y., Hsu, Y.Y., Wang, S.M., Liu, W.H., Tsai, C.H., Sheu, S.S., Chiang, P.C., Lin, W.P., Lin, C.H., Chen, W.S., Chen, F.T., Lien, C.H., Tsai, M.J.: Evidence and solution of over-reset problem for HfOX based resistive memory with sub-ns switching speed and high endurance. In: IEEE International Meeting on Electron Devices, pp. 19.7.1–19.7.4 (2010)

12. Shannon, C.E.: A symbolic analysis of relay and switching circuits. Thesis (M.S.)–Massachusetts Institute of Technology, Department of Electrical Engineering (1940)
13. Shirinzadeh, S., Soeken, M., Drechsler, R.: Multi-objective BDD optimization with evolutionary algorithms. In: Proceedings of the 2015 Annual Conference on Genetic and Evolutionary Computation (GECCO), pp. 751–758 (2015)
14. Shirinzadeh, S., Soeken, M., Drechsler, R.: Multi-objective BDD optimization for RRAM based circuit design. In: International Symposium on Design and Diagnostics of Electronic Circuits & Systems (2016)
15. Shirinzadeh, S., Soeken, M., Gaillardon, P.-E., Drechsler, R.: Fast logic synthesis for RRAM-based in-memory computing using majority-inverter graphs. In: Design, Automation & Test in Europe (2016)
16. Shirinzadeh, S., Soeken, M., Gaillardon, P., Micheli, G.D., Drechsler, R.: Endurance management for resistive logic-in-memory computing architectures. In: Design, Automation & Test in Europe, pp. 1092–1097 (2017)
17. Shirinzadeh, S., Soeken, M., Gaillardon, P.-E., Drechsler, R.: Logic synthesis for RRAM-based in-memory computing. IEEE Trans. Comput.-Aided Des. Integr. Circuits Syst. **37**, 1422–1435 (2017)
18. Soeken, M., Shirinzadeh, S., Gaillardon, P.-E., Amarù, L.G., Drechsler, R., De Micheli, G.: An MIG-based compiler for programmable logic-in-memory architectures. In: Design Automation Conference, pp. 117:1–117:6 (2016)
19. Soeken, M., Gaillardon, P.E., Shirinzadeh, S., Drechsler, R., Micheli, G.D.: A plim computer for the internet of things. Computer **50**(6), 35–40 (2017)
20. Wang, J., Dong, X., Xie, Y., Jouppi, N.P.: Endurance-aware cache line management for non-volatile caches. ACM Trans. Archit. Code Optim. **11**(1), 4:1–4:25 (2014)
21. Wong, H.S.P., Lee, H.Y., Yu, S., Chen, Y.S., Wu, Y., Chen, P.S., Lee, B., Chen, F.T., Tsai, M.J.: Metal-oxide RRAM. Proc. IEEE **100**(6), 1951–1970 (2012)
22. Xu, C., Niu, D., Muralimanohar, N., Balasubramonian, R., Zhang, T., Yu, S., Xie, Y.: Overcoming the challenges of crossbar resistive memory architectures. In: IEEE International Symposium on High Performance Computer Architecture, pp. 476–488 (2015)
23. Yang, S.: Logic synthesis and optimization benchmarks user guide: Version 3.0. MCNC (1991)
24. Zhang, H., Chen, G., Ooi, B.C., Tan, K.L., Zhang, M.: In-memory big data management and processing: a survey. IEEE Trans. Knowl. Data Eng. **27**(7), 1920–1948 (2015)

Chapter 4
Approximate Memory: Data Storage in the Context of Approximate Computing

Saman Froehlich, Daniel Große, and Rolf Drechsler

Abstract In modern computing many applications exist which do not require exact results or for which a single golden, correct solution does not exist. Consider web search as an example. The user does not know in what order which results would appear and considers any outcome as acceptable or correct for which the returned results are related to the search topic. These kinds of applications are inherently error tolerant. In particular, applications on mobile devices belong to this class, since they include games, digital media processing, web browsing, etc. However, in all of these applications, some data may be more resilient to errors than others. *Approximate Computing* (AC) is a design paradigm that tries to make use of this error tolerance by trading accuracy for performance.

In this chapter, we give an overview of techniques used to increase the performance of memory in modern circuits, by reducing the exactness requirements of the stored data (i.e., stored data may be lost or approximated by other stored data). After a short overview of the architecture of both SRAM and DRAM, we review techniques used for data partitioning. This is necessary in order to characterize which data may be stored approximately and which may not. Additionally, we describe techniques to reduce the power consumption as well as the performance of SRAM and DRAM. Finally, we present an approach that relaxes the requirement of exact matching when looking for a previously computed result to speed up the computation.

S. Froehlich (✉)
Institute of Computer Science, University of Bremen, Bremen, Germany
e-mail: froehlich@uni-bremen.de

D. Große · R. Drechsler
Group of Computer Architecture, Faculty 3 - Mathematics and Computer Science, University of Bremen, Bremen, Germany

Cyber-Physical Systems, DFKI GmbH, Bremen, Germany
e-mail: grosse@informatik.uni-bremen.de; drechsle@informatik.uni-bremen.de

C. S. Große, R. Drechsler (eds.), *Information Storage*,
https://doi.org/10.1007/978-3-030-19262-4_4

4.1 Introduction

We currently face an era in which small portable devices are growing ever more important. The *Internet of Things* (IoT), where all devices are connected to each other and can be controlled remotely, is a part of the everyday life in modern society. At the same time, the computational complexity of the applications which are executed on these devices increases, while the technology development is slowing down due to dark silicon (i.e., scaling the size of transistors has almost come to an end so that we can no longer implement more and more transistors on the same chip area and use them simultaneously due to power and thermal constraints).

However, many applications are inherently error tolerant. In particular, applications on mobile devices belong to this class, since they include games, digital media processing, web browsing, etc. [1]. Error tolerance is either caused by the limits of human perception (e.g., in digital audio and image processing) or due to the application itself (e.g., in web search or pattern recognition).

Approximate Computing (AC) is a design paradigm which tries to make use of this error tolerance by trading off accuracy for performance (i.e., computation time, circuit size/complexity, and/or power consumption). For the hardware side many approaches exist, ranging from functional approximation to voltage scaling. We focus on approximate memory in this chapter. The core concept is a redesign of the memory such that either the power consumption can be reduced (based, e.g., on decreasing the refresh rate of the memory) or the performance of the circuit can be increased. Besides altering the memory itself, the memory can be used as a look-up table for the purpose of approximation. For instance, the result of a computation can be stored as an input-output pair. If a computation with an input similar to that of a stored computation has to be calculated, the stored result can be loaded and used as an approximation for the true result of the computation.

The remainder of this chapter is structured as follows: In Sect. 4.2, we introduce the basic architecture of memory used in modern devices. We focus on SRAM (Static Random-Access Memory) and DRAM (Dynamic Random-Access Memory), since these are often subject to AC. We present techniques for data partitioning in Sect. 4.3. In Sect. 4.4 we describe techniques for approximation of SRAM and DRAM. We also explain how look-up tables are used for approximation of computations. Finally, Sect. 4.5 concludes the chapter.

4.2 Preliminaries

We first give an introduction to memories used in current computer architectures in Sect. 4.2.1. In Sect. 4.2.2 we introduce the basic structure of *Static Random-Access Memory* (SRAM), the structure of SRAM blocks, and their application in caches. In Sect. 4.2.3 we introduce the basic structure of *Dynamic Random-Access Memory* (DRAM), starting with a single DRAM cell up to the structure of whole DIMM blocks.

4.2.1 Introduction to Memory in Computer Architectures

In modern computer architecture, a large memory hierarchy is being used. In general, the faster the memory is, the more expensive it is. Thus, there are different memory layers, which increase in size but reduce in speed. The fastest memory are usually registers, followed by cache and main memory. The next layer (also known as secondary memory) are *Hard-Disk Drives* (HDDs) and *Solid-State Drives* (SSDs). This enumeration is by no means complete, but includes the most prominent representatives.

We will restrict ourselves to caches and the main memory, since these are often targeted by AC.

Caches and main memory are often implemented as volatile memory. In other words, the saved information is lost as soon as the supply voltage is being turned off. The two main technologies to implement volatile memory are SRAM, which uses transistors for data storage, and DRAM, which utilizes capacitors.

In the following we will first take a closer look at these technologies and summarize why they are interesting for AC.

4.2.2 Static Random-Access Memory (SRAM)

In this section we show the implementation of SRAM. In Sect. 4.2.2.1 we describe the structure of a single 6T-SRAM cell. In Sect. 4.2.2.2 we describe the structure of SRAM blocks. Thereafter, in Sect. 4.2.2.3 we discuss different cache types, where SRAM is heavily used.

4.2.2.1 SRAM Cells

An SRAM cell is used to store a *single bit of data*. It is depicted in Fig. 4.1. The *Word Line* (WL) denotes whether access to the SRAM cell is requested for both reading and writing operations. The *Bit Line* (BL) is used to read/write the actual value from/to the cell.

The transistors are ordered in such a way that if the BL is driven to V_{DD} or GND[1] and hence the $\overline{\text{BL}}$ to the inverse of BL, and the WL is set to 1, the transistors are switched to connect the BL to V_{DD} or GND, respectively. If the content of the cell is to be read, both BL and $\overline{\text{BL}}$ are set to 0.5 V_{DD}. Then, after setting the WL to 1, the BL is driven to the previously set value.

[1] V_{DD} is the common abbreviation for the *supply voltage* and GND for *ground*, respectively.

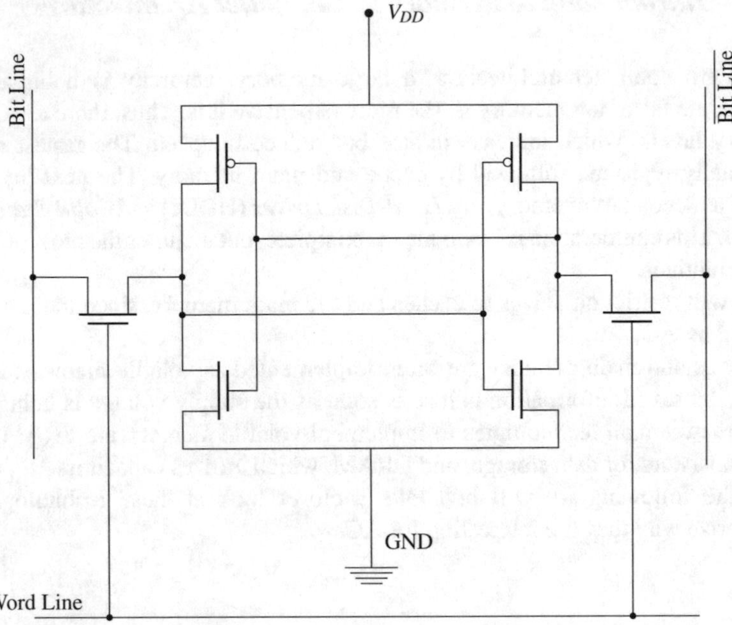

Fig. 4.1 SRAM cell

4.2.2.2 Structure of SRAM

In general, SRAM cells are ordered in blocks which store n-bit words. In low-power environments, additional sense amplifiers are attached to the BLs. This architecture is depicted in Fig. 4.2. The black lines shown in Fig. 4.2 represent the BLs. The blue lines are the WLs. During a read/write process, the address decoder activates the cells corresponding to the word which is to be accessed. Consecutively, the content of the corresponding cells is accessed using the BLs. To reduce the power consumption of the cells, the supply voltage V_{DD} is often kept at low levels. In order to be able to read values from the cells, a sense amplifier is needed.

The further the supply voltage V_{DD} is reduced, the more the resilience to influences of process, voltage, and temperature variations vanishes. Approaches exist to increase this resilience to allow for further reduction of the supply voltage (e.g., replica BLs, which mimic the behavior of the other BLs but with known values [2]).

4.2.2.3 Caches

A prominent application of SRAM are caches. A cache is a small but very fast memory, which is used to save recently used data. If this data is to be reused, it can be loaded directly from the cache without having to access slower memory.

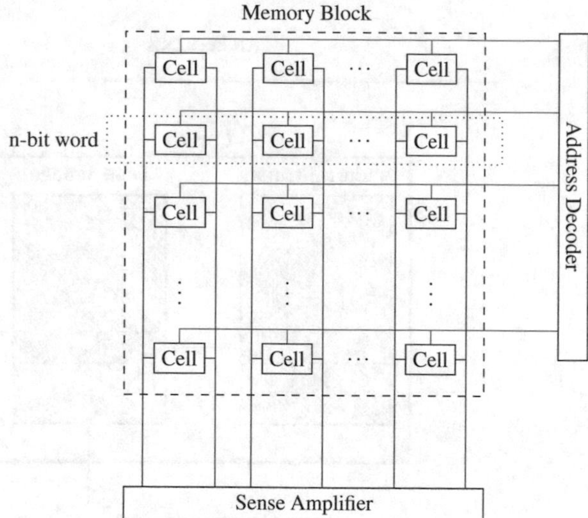

Fig. 4.2 SRAM structure

The idea for caches is based on the principle of locality. The principle of locality states that data, which is recently used, will be used again in the near future with high probability (*temporal locality*). It also states that if an address is accessed, the probability for an access to an adjacent memory address in the near future is high (*spatial locality*). This principle plays an important role in computer science.

The basic layout of a cache is depicted in Fig. 4.3. A cache consists of two parts: an address memory and a corresponding data storage. If data is to be read from the memory, the cache is checked before the main memory is accessed. It is checked whether the corresponding data is stored inside the cache. If so, the data is loaded. Otherwise, the main memory has to be accessed and the loaded data is stored inside the cache, so that when the data is to be loaded again, the cache can be used instead of the main memory.

There are different kinds of caches—the most fundamental ones are *direct mapped caches* and *full associative caches*. Direct mapped caches map a certain address range (usually indexed by the lower bits, due to spatial locality) to different cache entries. Every address range has a fixed entry, and if an entry already holds a value when the value of a new address is to be stored, the old one is simply overwritten. Due to their simplicity, direct mapped caches have little hardware overhead, which makes them the technology of choice if the cache has to be large.

An associative cache works in a different way. The cache entries are not tied to fixed address ranges. Their content can be set and overwritten dynamically. This results in a hardware overhead, but in return it can be more efficient. Also, for an associative cache, different strategies to replace existing data have to be considered, when new data has to be stored inside an already full cache. While in a direct mapped cache the old data is simply overwritten, different strategies exist for associative caches. The most prominent ones are to overwrite the *Least Frequently*

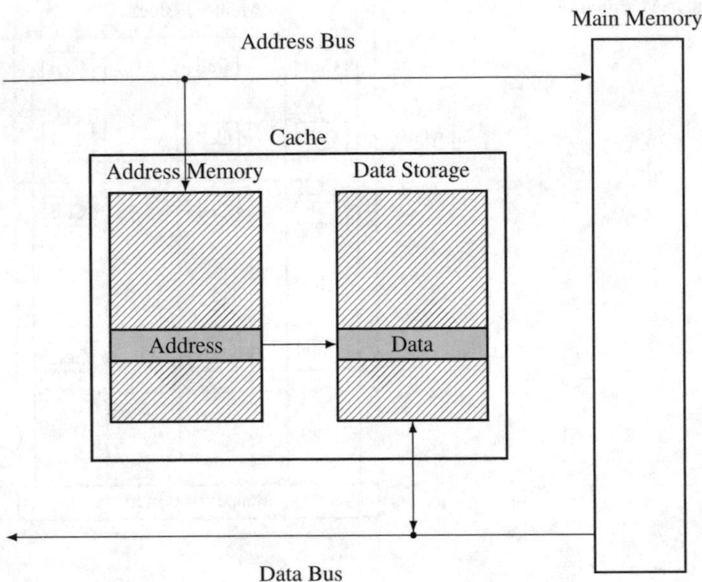

Fig. 4.3 Cache principle

Used (LFU), the *Least Recently Used* (LRU), or simply a *First-In, First-Out* (FIFO) strategy. In conclusion this means that while a full associative cache is more flexible, it requires a considerable hardware overhead. A direct mapped cache is less flexible, but requires less hardware overhead. For this reason, direct mapped caches are often larger than full associative caches.

In order to combine the advantage of both kinds of caches, hybrids exist. A prominent example is a four-way associative cache. This cache is based on a direct mapped cache, but instead of having a single entry for a certain address range, it has four entries (ways) which are each managed like a full associative cache. One could say that an n-way associative cache is like a direct mapped cache in which each cell is replaced with n-full associative caches.

4.2.3 Dynamic Random-Access Memory (DRAM)

In this section we present the structure of DRAM. We show the structure of single DRAM cells in Sect. 4.2.3.1. Subsequently, we describe the structure of DRAM blocks, which consist of many rows of DRAM cells in Sect. 4.2.3.2. Finally, in Sect. 4.2.3.3 we discuss the different operating modes of DRAM.

Fig. 4.4 DRAM cell

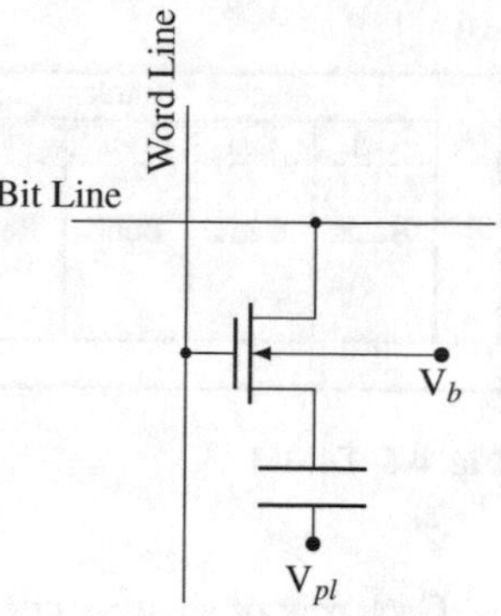

4.2.3.1 DRAM Cells

In contrast to the standard SRAM cell, a DRAM cell only consists of a single transistor (it is also known as single-transistor memory cell). In addition, a capacitor is included. A standard DRAM cell is depicted in Fig. 4.4.

In order to store a value into the DRAM cell, the desired value is loaded to the *Bit Line* (BL). After setting the *Word Line* (WL) to high, the value is stored into the capacitor. The lower part of the capacitor is not set to GND but to some voltage V_{pl} in order to reduce its charge. We are not going into more technical details here but refer the interested reader to other books such as [3]. The substrate of the transistor is also charged to an additional voltage V_b, for several reasons. However, again we are not going into further details.

After storing a value to the capacitor, its charge is slowly decreasing, mostly due to leakage of the transistor. For this reason a DRAM cell needs to be refreshed periodically.

The reading process is similar to that of an SRAM cell. The BL is charged to $\frac{1}{2}$, and after setting WL to 1, the charge of the capacitor will adjust the charge of the BL accordingly.

4.2.3.2 Structure of DRAM

In general-purpose computers, DRAM is often separated into several *Dual In-line Memory Modules* (DIMMs). A DIMM consists of one or more *Ranks*, each of which consists of several Banks. A *Bank* is a two-dimensional array, which finally holds the cells [4]. Thus the hierarchy is:

- DIMM
- Rank
- Bank
- Cell

This hierarchy is also depicted in Fig. 4.5.

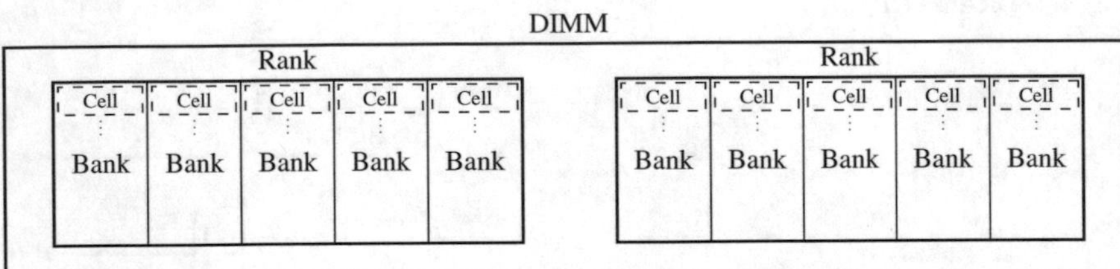

Fig. 4.5 DIMM

Each row of an array corresponds to one WL to which many cells are connected. The columns are the BLs to which again many cells are connected. In order to store or read data to/from DRAM memory, a certain WL is addressed. The corresponding WL is set to 1 and thus the BLs are loaded with the stored charge of the corresponding cells. Just like in SRAM, additional sense amplifiers may be used (not depicted).

4.2.3.3 Operation Modes of DRAM

DRAM operates in different modes, depending on the state of the system [1]. We give a simplified description:

Activate/precharge	*Activate* and *precharge* are the active states in which the DRAM can be accessed for read and write operations.
Fast low-power	The *fast low-power mode* is used during idle states. This mode has a short wake-up time and thus gives a good power/performance trade-off when programs are executed, which need light access to the DRAM regularly.
Self-refresh/deep power-down	*Self-refresh* and *deep power-down are modes* which are used when the system is on standby. They need time to power back up to full functionality, but result in huge power savings. In contrast to self-refresh, the mode deep power-down will stop all refresh operations and thus the saved information is gradually fading.
Partial array self-refresh (PASR)	*Partial array self-refresh* (PASR) is a hybrid between the self-refresh and the deep power-down mode. In PASR mode, only parts of the DRAM are refreshed. The OS is responsible for specifying which parts of the DRAM are being refreshed. Data in Banks which are not refreshed is discarded.

4.2.4 SRAM vs. DRAM

We have now introduced SRAM and DRAM. Before we get to the application of AC in memory, we want to compare these technologies to each other and finally get to their use case in common applications.

In general, SRAM has a better performance when it comes to power consumption and speed. Unlike DRAM, SRAM does not need to be refreshed periodically in order to guarantee that the stored data is not lost. In the SRAM technology, the transistors which store the data hold themselves.[2] Since SRAM does not need to load a capacitor to a certain value, it is also superior when it comes to speed.

DRAM on the other hand is a lot cheaper than SRAM. In comparison, a standard SRAM cell needs six transistors (though other implementations with less transistors exist), while DRAM only needs one transistor and one capacitor.

This is why DRAM is usually used for the main memory, which is larger and slower, while SRAM is used for cache memory, which is often smaller than the main memory, but needs to have shorter access times. SRAM is also important for low-power devices, since the refresh rate of DRAM makes it too power consuming for many applications.

While SRAM is traditionally used in applications like mobile phones, the increasing demand on memory in such devices leads to deployment of DRAMs. Thus, techniques for the approximation of both SRAM and DRAM are of interest.

4.3 Data Partitioning

In many applications there is data which may be approximated and data which must not be approximated. The user of a system that supports AC usually knows which part may be subject to approximations and which may not.

EnerJ [5] is an extension for the programing language Java, which supports annotation of parts that may be approximated. It also provides data types that may be subject to approximation (either due to approximate operations and functions or due to approximate memory). Its main purpose is to provide a portable way for implementing approximate programs, independent of hardware and approximation miter.

EnerJ allows the definition of approximate variables using the annotations. Member variables of classes can be defined depending on whether the object itself is instantiated as approximate. Precise variables are declared either implicitly by not using any annotation or explicitly. Member functions can have approximate overloads. Instead of an annotation, the function needs to have a special prefix.

[2]This means that they keep their state until another value is stored.

As an example, the approximate overload of a function *mean()* would be named *mean_APPROX()*.

Precise and approximate variables are incompatible in general. In practice however, it may be necessary to assign the value of an approximate variable to a precise one. EnerJ provides a function called *endorse* to do so. So if p denotes a precise and a denotes an approximate variable of the same type, then $p=a$ would be invalid, while $p=endorse(a)$ would run perfectly fine. This protects the programmer from mixing up variables. The assignment of precise values to approximate variables can be executed without further aid.

There are some restrictions to approximate variables. They may not be used as conditions to statements that affect the control flow (such as if and while clauses). In order to prevent illegal memory access, it further prohibits the use of approximate variables to access elements stored in arrays. These restrictions may be bypassed by using the *endorse* statement.

Since EnerJ is not sound (which usually is not a problem in practice according to the authors), the authors have also introduced FEnerJ which is based on Featherweight Java. FEnerJ is sound in a way that it strictly separates approximate and precise data [5].

EnerJ relies on the underlying hardware to provide support for approximations. The authors of [6] have implemented a processor which includes a look-up table to store intermediate results of floating-point multiplications. The programmer can configure how well the inputs to a multiplication must match the inputs of the stored results of the look-up table to be evaluated as a match. The interface is provided as a C++ library. The underlying hardware and experimental results are presented in more detail in Sect. 4.4.3.1. Processors with a similar functionality have been proposed in [7, 8].

4.4 Approximate Memory

The straightforward way is to store data approximately, for example, in case of images. In order to mask manufacturing variations, memory is in general "overdesigned" using higher supply voltages for SRAM (guard banding) [9, 10] and higher refresh rates than required for DRAM [9, 10]. If the application is error tolerant, these safety margins can be reduced tremendously and thus AC is a promising paradigm to use. An alternative is to relax the requirement of data matching when looking for a previously computed result to speed up the computation [6, 11].

In Sect. 4.4.1 we describe techniques employed for the implementation of approximate SRAM. Consecutively, we describe approximate DRAM in Sect. 4.4.2. Section 4.4.3 presents an approach for relaxed data matching.

4.4.1 Approximate SRAM

SRAM is heavily used in embedded and mobile devices. It also occupies most of the area in *Very-Large-Scale Integration* (VLSI) systems, and its speed and power consumption have a notable impact on their performance [12]. These devices often have constraints when it comes to power consumption, and thus low power is an important design goal. For this reason approximate SRAM aims at reducing the power consumption of the memory. In contrast to DRAM, SRAM does not need to be refreshed due to its architecture. The largest part of the power it consumes is caused by leakage currents. Voltage scaling is one technique employed in this process (e.g., [2, 9, 10, 12, 13]). This technique has proven to be effective, since the dynamic power dissipation depends quadratically on the supply voltage V_{DD}. However, failure due to noise in SRAM heavily increases if the supply voltage is reduced, since the resilience toward process, voltage, and temperature variation decreases [2].

4.4.1.1 Relaxed Cache: Relaxing Manufacturing Guard Bands in Memories for Energy Saving

Functionality

The authors of [9, 10] have introduced an approximate cache. Caches are among the most common applications for SRAM. In [9, 10], cache with a fixed number of ways, each of which consists of a fixed number of blocks, is considered. Some ways are marked as reliable (i.e., with a fixed supply voltage, such that the stored data is guaranteed to be fault-free under normal conditions). These are called *protected ways*. Others are so-called relaxed ways, which means that they have a relaxed correctness requirement. The supply voltage for the relaxed ways can be scaled. The authors propose two design parameters: supply voltage (V_{DD}) and the number of *Acceptable Faulty Bits* (AFB). Whenever an application is designed, the programmer can set these design parameters and an acceptable SRAM block is selected.

To be able to identify suited cache blocks for a (V_{DD}, AFB) combination, different SRAM blocks have to be characterized. In order to reduce hardware overhead, the number of different supply voltages for the relaxed ways is fixed (i.e., if the standard supply voltage is 600 mV, there may be four levels for reduced supply voltage $V1 = 560$ mV, $V2 = 520$ mV, $V3 = 480$ mV, $V4 = 440$ mV). The number of possible AFB levels is also fixed (i.e., there may be four levels with $AFB1 = 1$, $AFB2 = 2$, $AFB3 = 3$, $AFB4 = 4$). *Built-In Self-Test* (BIST) routines exist to detect memory faults (e.g., the March test). These are applied to each block at different supply voltages to determine the number of faulty bits. Such a BIST can either be only executed once for the whole device lifetime or repeatedly at the start-up. If the BIST is only run once, the results must be stored in a nonvolatile memory, which results in an additional hardware overhead. Otherwise, no additional nonvolatile memory is required. These results (defect map) are used to characterize which supply voltage for a memory block is needed, in order to guarantee a given maximum number of faulty bits.

In general, the memory blocks can be divided into four categories:

Protected Blocks — *Protected Blocks* are part of the protected ways and voltage scaling is not applied. They can be used to store critical data safely.

Clean Blocks — Eventhough *Clean Blocks* are part of the relaxed ways, the current level of voltage scaling does not cause a reduced performance in terms of reliability.

Relaxed Blocks — *Relaxed Blocks* are subject to voltage scaling. The current level of voltage scaling does cause some unreliability, but not to a level where the limit for the AFB is violated.

Disabled Blocks — The current level of voltage scaling does cause an unreliability up to a level for which integrity in terms of the currently set level of AFB cannot be guaranteed. *Disabled Blocks* cannot be used to store information unless the supply voltage or the limit for AFB is increased.

Figure 4.6 demonstrates these terminologies for a four-way SRAM with one protected way and scaled voltage. The number inside each block represents the number of faulty bits, respectively. The AFB has been set to 3, so any block with more than 3 faulty bits is disabled (marked in red).

Evaluation

In order to evaluate their approach, the authors of [9, 10] have altered the gem5 framework [14] for implementation and applied it to several benchmarks. The basic of these benchmarks is *Susan*, an image recognition package from MiBench [15].

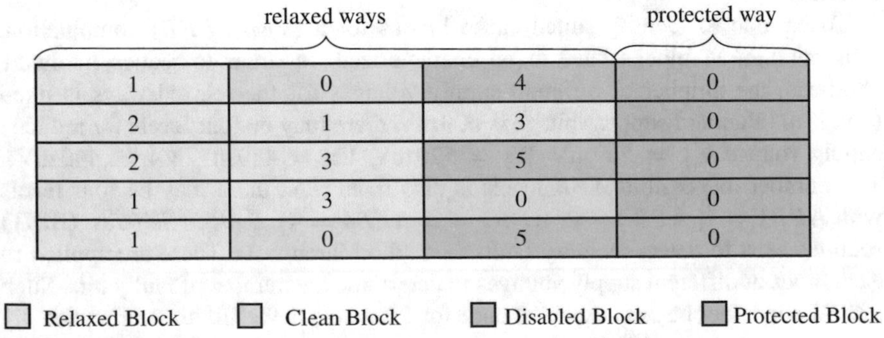

relaxed ways			protected way
1	0	4	0
2	1	3	0
2	3	5	0
1	3	0	0
1	0	5	0

Relaxed Block   Clean Block   Disabled Block   Protected Block

Fig. 4.6 Example: SRAM cache with four ways, reduced V_{DD}, and $AFB = 3$

Susan has been used for image scaling, image smoothing, edge detection, and video encoding. The cache used for benchmarks is assumed to run at a baseline supply voltage of $V_{DD} = 600\,\text{mV}$. The voltage has been scaled to 560, 520, 480, and 440 mV.

The experiments can be summarized as follows:

Image scaling: *Peak Signal-to-Noise Ratio* (PSNR) is an established quality metric in the image processing domain. High PSNR values correspond to a good perceptual quality, while low values correspond to a noisy picture. It has been used in [9, 10] to measure the effects of the presented approach on the quality of the result of the image scaling benchmarks. Pictures with PSNR$\geq$ 28.0 dB are said to be acceptable; pictures with PSNR above 30.0 dB are said to have a good quality. AFB is varied between 1 and 4.

The results presented in [9, 10] show that the voltage can be scaled down to 440 mV and $AFB = 3$ and still generate acceptable results. The increase in computation time due to deactivated cache blocks is less than 3%.

Image smoothing: For image smoothing, PSNR has again been used as a quality measure and AFB has been varied between 1 and 4. The results reported in [9, 10] are very promising. For $V_{DD} \geq 520\,\text{mV}$, any AFB between 1 and 4 results in an at least acceptable outcome. For $V_{DD} < 520\,\text{mV}$, still acceptable AFB values exist, such that voltage can be scaled aggressively. The average PSNR is higher than that of the image scaling benchmark. Apparently the image smoothing application allows for more aggressive voltage scaling than image scaling. This confirms the assumption that different applications have different error resilience and thus V_{DD} and AFB should be adopted dynamically. The increase in computation time due to deactivated memory cells is less than 2%.

Edge detection: In order to measure the quality of the edge detection result, the ratio between true and false positives is measured. A ratio of 0.8 is considered acceptable. Shoushtari et al. [9, 10] report that for $V_{DD} = 480\,\text{mV}$, an AFB of 6 still yields acceptable results. Apparently the edge detection application is even more error resilient than image scaling and image smoothing. The increase in computation time due to deactivated memory cells is less than 3%.

Video encoding: The video encoding application is more difficult to evaluate than the other three. Video encoding is a memory hungry application. Reducing V_{DD} without increasing AFB disables cache cells. This results in more cache misses, which in turn leads to a more frequent access to the main memory. The consequence is significant reductions in performance and decoded *Frames Per Second* (FPS) (an FPS loss of more than 35% has been reported for $V_{DD} = 460\,\text{mV}$ and $AFB = 4$). However, simply increasing AFB can lead to degenerated results. In general, a trade-off between energy savings, FPS, and output quality has to be considered. It is up to the programmer to set applicable values, since sometimes an early result with bad quality may be more desirable than a late result with high quality. However, in all experiments the PSNR of the output was still above 32 dB.

We have now summed up the effects of V_{DD} and AFB to different applications. We now explain the influence on power consumption. Shoushtari et al. [9, 10] assume power gating (which means that disabled cells are powered off, resulting in a reduction in leakage energy). If the supply voltage is kept constant but AFB increases, disabled memory is reclaimed and saving in leakage energy decreases. To evaluate the power savings in [9, 10], a baseline supply voltage of 700 mV is assumed and scaled down to 440 mV stepwise. It appears that for $V_{DD} \geq 540$, increasing AFB beyond 1 does not reduce the savings in leakage energy. Apparently, the number of faulty blocks is very small at this voltage. The analysis in [9, 10] shows that leakage energy can be reduced by up to 74%, depending on the configuration of V_{DD} and AFB.

In conclusion, the benefit of the presented approach strongly depends on the application. For applications with moderate memory usage (image smoothing, edge detection, image scaling), a significant reduction of power consumption depending on $(V_{DD},\ AFB)$ has been reported. However, for applications that use a large amount of traffic, only very little gain is reported, and wrong configuration can even lead to an immense increase in computation time. This is due to the reduced cache size, because of deactivated blocks. For this kind of applications $(V_{DD},\ AFB)$ has to be chosen very carefully.

4.4.2 Approximate DRAM

As stated earlier, DRAM needs to be refreshed regularly in order to guarantee that the stored data is not compromised (see Sect. 4.2.3). A refresh interval of 64 ms is a common value for commercial DRAM. It has been predicted that the refreshing procedure will cause 40–50% of the overall power consumption in future devices [16, 17]. Thus, many approaches exist for which the refresh rate is being reduced in order to save energy [18].

While a regular refresh rate is imperative for correctness, reducing the refresh rate comes with two advantages: The first advantage is the reduction in power consumption. The second advantage is a reduction of access time: While a cell is being refreshed, it cannot be accessed for any operations [4, 18]. In general a DRAM DIMM is not refreshed cell by cell but Rank by Rank. During such a refresh command, all the Banks within a Rank are not available for access, and the reading/writing process needs to wait for the refresh operation to finish.

The authors of [18] have shown that reducing the refresh rate can cause unpredictable bit flips in the stored data and thus a non-deterministic degradation of the output. This is due to small differences between the cells (caused by the manufacturing). Also, the necessary refresh rate depends on the temperature and other factors [19].

It is difficult to characterize the error behavior of DRAM in an analytic approach, due to the variations between cells (caused by the manufacturing process), their variating resilience to changes in temperature, and the fact that the retention time of a cell also depends on the stored data pattern. So either stochastic models are being used or representative reference data has to be available.

In order to allow for data partitioning (see Sect. 4.3), the refresh rate is often not uniform across DRAM, but varies between DIMMs and/or Banks, such that critical data can be stored in cells which are refreshed more often compared to cells for non-critical data (e.g., [1]).

4.4.2.1 Flikker

Functionality

Flikker, introduced in [1], was the first approach toward a software-based control of approximate DRAM. The basic idea behind Flikker is to lower the refresh rate (and thus its reliability) for parts of the DRAM Banks. In this context, critical data can be saved in the parts of DRAM which is refreshed at the regular refresh rate T_{regular}, while other data can be stored in the less reliable part of the memory (with the lower refresh rate T_{low}). The main difference between Flikker and the *partial array self-refresh* (PASR) mode (see Sect. 4.2.3.3) is that while in PASR the data in Banks which are not refreshed gets lost, in Flikker the reliability of the data is lowered.

Flikker requires small changes in the hardware architecture, as it requires to configure the refresh rate of DRAM online, controlled by software. Existing DRAM architecture must be modified in such a way that it allows for different refresh rates of different sections. Mobile DRAM uses a hardware counter while in self-refresh mode to remember which row to refresh next. Flicker adds extra bits [1] to vary the refresh rate between rows:

- The counter is extended by some additional bits, depending on the ratio of T_{low} to T_{regular}. If, for example, $\frac{T_{\text{low}}}{T_{\text{regular}}} = \frac{1}{16}$, then four additional bits are added to the counter (such that the counter can count up to 16). Each time the last row is refreshed, the counter is increased by 1.
- The extended counter has a *refresh enable bit*. The current row is only refreshed, if the refresh enable bit is set to 1.

The refresh enable bit is set by a configurable controller, depending on the higher-order bits of the address of the current row and the combination of the other additional bits. If the address belongs to parts of the DRAM, which are refreshed at T_{regular}, the refresh enable bit is always set to 1. Otherwise, it is set depending on the other additional counter bits. In this implementation, the Banks in DRAM are always partitioned in such a way that the reliable part of the DRAM has the same higher-order bits (and thus is at the beginning or the end of the DRAM).

Evaluation

The efficiency of this approach is subject to some parameters:

1. The resilience toward variations of the refresh rate of a cell depends on the temperature. The authors of [1] have assumed an operating temperature of 48 °C, since they assume that this is above the operating temperature of most mobile devices and thus it will perform better in everyday use.
2. The efficiency depends on the size of portion of the DRAM, which may be substitute to a reduced refresh rate.
3. The lower T_{low} is, the more energy is saved and the higher the error rate will be.

The model for power consumption provided in [1] is based on power measurements of real DRAM in PASR mode. It is given as:

$$P_{\text{Flikker}} = P_{\text{refresh}} + P_{\text{other}} \tag{4.1}$$

$$P_{\text{Flikker}} = P_{\text{refreshlow}} + P_{\text{refreshhigh}} + P_{\text{Other}} \tag{4.2}$$

$$P_{\text{Flikker}} = (P_L)\frac{T_{\text{regular}}}{T_{\text{low}}} + P_{\text{refreshhigh}} + P_{\text{Other}} \tag{4.3}$$

$$P_{\text{Flikker}} = (P_{\text{full}} - P_{\text{PASR}})\frac{T_{\text{regular}}}{T_{\text{low}}} + P_{\text{PASR}} \tag{4.4}$$

The overall power consumption can be separated into the sum of the power consumed by the refresh of the DRAM cells and other components (i.e., control logic and the like). This is represented by Eq. (4.1). The power consumption of the refresh can be split into that of the parts of the DRAM with high refresh rate $P_{\text{refreshhigh}}$ and that of the parts with low refresh rate $P_{\text{refreshlow}}$ (Eq. (4.2)). $P_{\text{refreshlow}}$ can be represented as the product of some constant P_L with the ratio of T_{regular} and T_{low}, as shown in Eq. (4.3). If T_{low} is very large (i.e., approximately ∞), the power consumption of Flikker is the same as that of RAM in PASR mode P_{PASR}. On the other hand, if T_{low} is the same as T_{refresh}, the power consumption of Flikker is the same as that of DRAM in full operation mode P_{full}. This is given in Eq. (4.4).

In order to calculate the trade-off between power and error rate for Flikker, besides a model for the power consumption, a model for the error rate is needed. The authors of [1] propose to use the results of [20, 21], since they have measured

Table 4.1 Error rate for different refresh cycles at 48 °C

Refresh cycle [s]	Error rate	Bit flips per byte
1	4.0×10^{-8}	3.2×10^{-7}
2	2.6×10^{-7}	2.1×10^{-6}
5	3.8×10^{-6}	3.0×10^{-5}
10	2.0×10^{-5}	1.6×10^{-4}
20	1.3×10^{-4}	1.0×10^{-3}.

error rate as a function of refresh rate. Different error rates were derived in [1] from the data given in [20]. These are given in Table 4.1.

From the previous analysis in error rate and power consumption, the trade-off between power consumption and error rate can be analyzed. In [1] it has been shown that the increase in error rate is exponential, while the saving in power is saturated at some point, if T_{low} is increasing. For the technology used in [1], a good value for T_{low} is $1s$, since the error rate keeps increasing, while the power saving has almost reached its final value (e.g., increasing T_{low} to 20 s would result in the increase of power saving from 22.5 to 23.9%, while the error rate is increased by a factor of more than 3000). If T_{low} is reduced to $0.5s$, the power saving is reduced tremendously.

4.4.3 *Approximate Computing by Relaxed Data Matching*

Another way to use memory for AC is to relax the requirement of data matching when looking for previously computed results to speed up computations. The inputs of a computation are used as a key to find a corresponding cell in a cache memory in which the corresponding result is stored. If used for AC, the strict requirement of the input of the computation to match the inputs of the stored result can be relaxed.

4.4.3.1 ProACt

Functionality

In [6] we present ProACt, a *Processor for high-performance on-demand Approximate Computing* which is designed especially for AC. The core idea of ProACt is to functionally approximate floating-point operations using previously computed results from a cache, thereby relaxing the requirement of having the exact identical input values for the current floating-point operation. To enable on-demand approximation, we add a custom instruction to the *Instruction Set Architecture* (ISA) of the used processor which essentially adjusts the input data for cache look-up and by this controls the approximation behavior. Overall a ProACt development

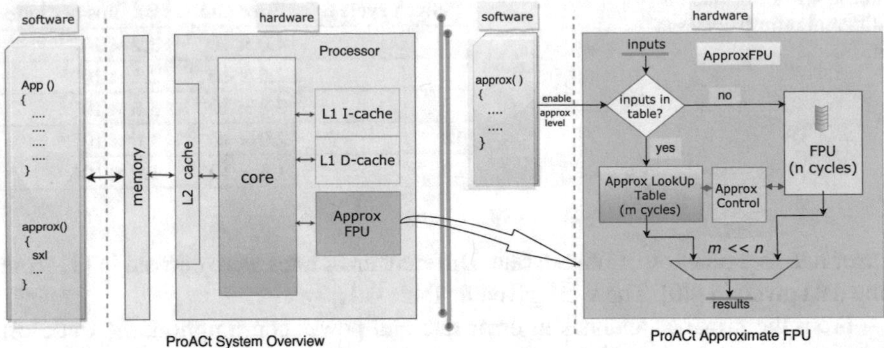

Fig. 4.7 ProACt system overview [6]

framework is devised. This framework consists of an extended hardware processor[3] and a software tool chain (cross-compiler, linker, etc.) to build a complete system for on-demand AC.

The ProACt system overview is shown in Fig. 4.7. As can be seen, it consists of the processor hardware and the software units working together to achieve approximation in computations. To operate the approximations in hardware, the *Approximate Floating-Point Unit* (AFPU) is added (a zoom is given on the right hand side of Fig. 4.7). In normal mode (i.e., approximations disabled), ProACt floating-point results are IEEE-754 compliant.

The AFPU in ProACt consists of an approximation look-up table, a pipelined *Floating-Point Unit* (FPU), and an approximation control logic (see Fig. 4.7, right hand side). The central approach used is that the results of the FPU are stored inside the look-up table first, and further operations are checked in this look-up table, before invoking the FPU for subsequent computation. The input arguments to the FPU are checked in the look-up table, and when a match is found, the results from the table are fed to the output, bypassing the entire FPU. The FPU will process only those operations which do not have results in the table. This look-up mechanism is much faster, resulting in significant savings in clock cycles. Approximation masks are applied to the operands before checking the look-up table. Thus, the accuracy of the results can be traded off using these masks. These approximation masks are set by the software (via a custom approximation control instruction) and vary in precision. The mask value (or alternatively called *approximation level*) denotes the number of bits to be masked from the LSB, before checking for an entry in the look-up table.

The software compiler relies on the ISA to transform a program to a binary executable. Hence, the ISA is extended with a single assembly instruction `SXL` (*Set approXimation Level*) for the software control of approximations. SLX is designed

[3]ProACt is based on RISC-V [22], a state-of-the-art open-source 64-bit RISC architecture.

as an immediate instruction that takes an 11-bit immediate value. The LSB, when set to 1, enables the hardware approximations. The remaining bits are used to set the approximation level and other special flags. The software can also disable the approximation unit by setting the level as 0. Here, it will simply act as a result caching and look-up mechanism without any approximation.

Evaluation

In [6], two different applications were used to evaluate ProACt: The first one is an image processing application and the second set consists of mathematical functions from scientific computing.

Table 4.2 shows the results from a case study on edge detection [23] using ProACt. The top row of images (Set 1) in Table 4.2 is generated with approximations disabled by the supervisory program. The middle row (Set 2) is generated with approximations enabled, and the last row shows bar plots for the hardware cycles taken by the core algorithm, along with the speedup obtained.

Table 4.2 Edge detection with approximations [6]

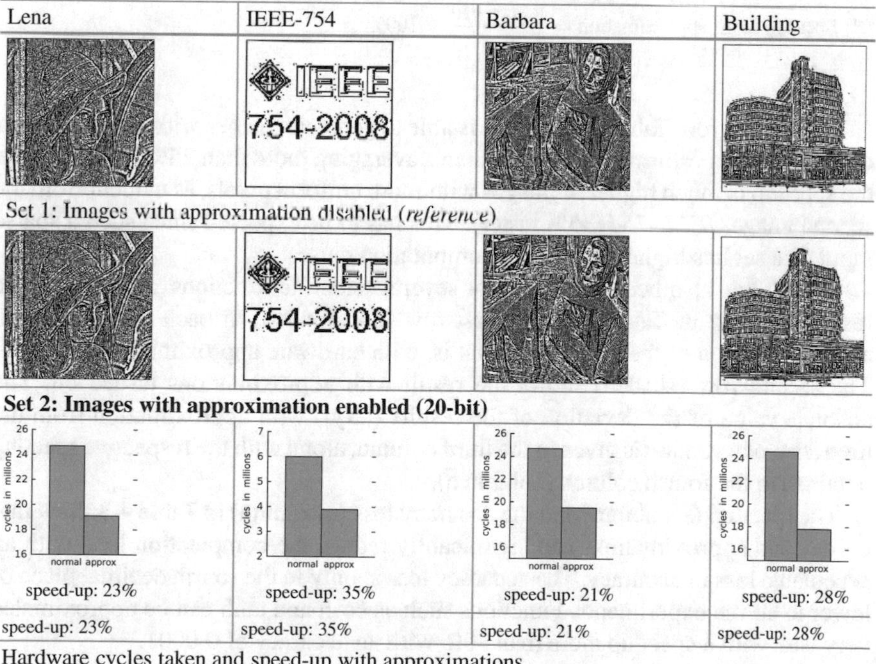

Lena	IEEE-754	Barbara	Building
Set 1: Images with approximation disabled (*reference*)			
Set 2: Images with approximation enabled (20-bit)			
speed-up: 23%	speed-up: 35%	speed-up: 21%	speed-up: 28%
Hardware cycles taken and speed-up with approximations			

Images generated from ProACt FPGA hardware
Set 1 reference images are with normal processing
Set 2 images are with approximation enabled (20-bit)

Table 4.3 Math functions with approximations [6]

Appx level (*a*)	Cycles n (*b*)	$\|\Delta y\|$ $\times 10^{-3}$ (*c*)	Speed up% (*d*)	Appx level (*a*)	Cycles n (*b*)	$\|\Delta y\|$ $\times 10^{-3}$ (*c*)	Speed up% (*d*)
$y = \sinh(x)$				$y = \sinh^{-1}(x)$			
−1	11083	0.00	0.00	−1	76,899	0.00	0.00
20	**7791**	**0.15**	**29.70**	**20**	**72,506**	**3.91**	**5.71**
$y = \cosh(x)$				$y = \cosh^{-1}(x)$			
−1	10,820	0.00	0.00	−1	78,616	0.00	0.00
20	**7501**	**0.14**	**30.67**	**20**	**73,843**	**2.14**	**6.07**
$y = \tanh(x)$				$y = \tanh^{-1}(x)$			
−1	10848	0.00	0.00	−1	7698	0.00	0.00
20	**7505**	**0.10**	**30.82**	**20**	**6135**	**0.93**	**20.30**

Functions $y = f(x)$ evaluated in ProACt FPGA hardware
Bold values are results with approximations enabled
(*a*) Approximation level (set with `SXL` instruction)
−1: high accurate result (approximation fully disabled)
20: 20-bit approximation in float division
(*b*) Number of machine cycles (n) taken for computation
(*c*) Accuracy, $|\Delta y| = |y_{-1} - y_{20}| \times 10^{-3}$
(*d*) Speedup from approximation $= \dfrac{n_{-1} - n_{20}}{n_{-1}} \times 100\%$

As evident from Table 4.2, ProACt is able to generate images with negligible loss of quality with performance improvements averaging more than 25%. Furthermore, the speedup is much higher in images with more uniform pixels, as evident from the second image, *IEEE-754* (35% faster). This has to be expected since such a sparse input data set has higher chances of computation reuse.

ProACt has also been evaluated for several scientific functions. A subset of the results is given in Table 4.3. The first row (non-shaded) in each set is the high accuracy version of the algorithms, that is, with hardware approximations disabled. The second row (shaded) shows the result with approximations turned on. The absolute value of the deviation of the results ($|\Delta y|$) with approximation from the high accuracy version is given in the third column, along with the respective speedup obtained in the fourth column (column d).

The speedup (*d* column) and the accuracy loss (*c* column) in Table 4.3 show that on-demand approximations can significantly reduce the computation load with an acceptable loss in accuracy. The accuracy loss is only in the fourth decimal place or lower in all the experiments. Functions such as cosh and tanh can be approximated very well with a speedup more than 30% with an accuracy of 0.0001.

4.5 Conclusion

Approximate Computing is a promising design paradigm, which allows to use resources a lot more effective than conservative computing for many applications.

In this chapter we have given an example for the techniques used in AC by surveying memory-based techniques. First, we have reviewed the basic functionality of SRAM and DRAM memory in Sect. 4.2. For SRAM, we have introduced the structure of SRAM blocks and their application to cache. For DRAM, we have introduced the structure of DIMM and the different operating modes.

We have provided insight in techniques to partition data in Sect. 4.3 and presented AC techniques for both SRAM and DRAM in Sect. 4.4. The technique for SRAM is based on voltage scaling (reducing the supply voltage). It has been applied to several applications and it could be shown that significant gains are possible. However, the margin of the gain is very application dependent. The technique for DRAM is based on the reduction of the refresh rate. Again, notable speedups at a reasonable error rate become possible.

Finally, we have introduced an approach that relaxes the requirement of exact matching when looking for a previously computed result to speed up the computation. The approach has been realized as a processor for high-performance on-demand AC and provides an interface to set the approximation level online. It has been evaluated for edge detection and the computation of mathematical functions. Again, notable speedups have been achieved.

In future work it seems promising to integrate approximate memory into ProACt. Furthermore, the applicability of formal methods as, for example, proposed in [24–26] for error estimation and synthesis/hardware generation, should be investigated. Designing application-specific memory from a correct-by-construction perspective is another line of research; see, for instance, [27].

Acknowledgements The authors would like to thank Arun Chandrasekharan for his significant contributions to ProACt.

This work was supported in part by the German Research Foundation (DFG) within the project MANIAC (DR 287/29-1), by the Reinhart Koselleck Project DR 287/23-1, and by the University of Bremen's graduate school SyDe, funded by the German Excellence Initiative.

References

1. Liu, S., Pattabiraman, K., Moscibroda, T., Zorn, B.G.: Flikker: saving dram refresh-power through critical data partitioning. SIGPLAN Not. **46**(3), 213–224 (2011)
2. Ataei, S., Stine, J.E.: A 64 kb approximate sram architecture for low-power video applications. IEEE Embed. Syst. Lett. **10**, 10–13 (2017)
3. Baker, R.J.: CMOS Circuit Design, Layout, and Simulation, 3rd edn. Wiley-IEEE Press, Hoboken (2010)
4. Lui, J., Jaiyen, B., Kim, Y., Wilkerson, C., Mutlu, O.: An experimental study of data retention behavior in modern dram devices: implications for retention time profiling mechanisms. In: Proceedings of the 40th Annual International Symposium on Computer Architecture (ISCA), pp. 60–71 (2013)

5. Sampson, A., Dietl, W., Fortuna, E., Gnanapragasam, D., Ceze, L., Grossman, D.: EnerJ: approximate data types for safe and general low-power computation. In: Proceedings of the 32nd ACM SIGPLAN Conference on Programming Language Design and Implementation (PLDI), pp. 164–174 (2011)
6. Chandrasekharan, A., Große, D., Drechsler, R.: ProACt: a processor for high performance on-demand approximate computing. In: ACM Great Lakes Symposium on VLSI, pp. 463–466 (2017)
7. Esmaeilzadeh, H., Sampson, A., Ceze, L., Burger, D.: Architecture support for disciplined approximate programming. In: International Conference on Architectural Support for Programming Languages and Operating Systems, pp. 301–312 (2012)
8. Venkataramani, S., Chippa, V.K., Chakradhar, S.T., Roy, K., Raghunathan, A.: Quality programmable vector processors for approximate computing. In: 46th Annual IEEE/ACM International Symposium on Microarchitecture (MICRO), pp. 1–12 (2013)
9. Shoushtari, M., Banaiyan, A., Dutt, N.: Relaxed cache: relaxing manufacturing guard-bands in memories for energy saving. Center for Embedded Computer Systems, Irvine, CA, Tech. Rep. (2014)
10. Shoushtari, M., Banaiyan, A., Dutt, N.: Exploiting partially-forgetful memories for approximate computing. IEEE Embed. Syst. Lett. **7**(1), 19–22 (2015)
11. Imani, M., Rahimi, A., Rosing, T.S.: Resistive configurable associative memory for approximate computing. In: Design, Automation and Test in Europe, pp. 1327–1332 (2016)
12. Ataei, S., Stine, J.E.: Multi bitline replica delay technique for variation tolerant timing of SRAM. In: ACM Great Lakes Symposium on VLSI, pp. 173–178 (2015)
13. Yang, L., Murmann, B.: Approximate sram for energy-efficient, privacy-preserving convolutional neural networks. In: IEEE Annual Symposium on VLSI, pp. 689–694 (2017)
14. Binkert, N., Beckmann, B., Black, G., Reinhardt, S.K., Saidi, A., Basu, A., Hestness, J., Hower, D.R., Krishna, T., Sardashti, S., Sen, R., Sewell, K., Shoaib, M., Vaish, N., Hill, M.D., Wood, D.A.: The gem5 simulator. SIGARCH Comput. Archit. News **39**(2), 1–7 (2011)
15. Guthaus, M.R., Ringenberg, J.S., Ernst, D., Austin, T.M., Mudge, T., Brown, R.B.: Mibench: a free, commercially representative embedded benchmark suite. In: 2001 IEEE International Workshop on Proceedings of the Workload Characterization, 2001. WWC-4, pp. 3–14 (2001)
16. Bhati, I., Chishti, Z., Lu, S.-L., Jacob, B.: Flexible auto-refresh: enabling scalable and energy-efficient dram refresh reductions. In: International Symposium on Computer Architecture (ISCA), pp. 235–246 (2015)
17. Liu, J., Jaiyen, B., Veras, R., Mutlu, O.: RAIDR: retention-aware intelligent dram refresh. In: International Symposium on Computer Architecture (ISCA), pp. 1–12 (2012)
18. Raha, A., Sutar, S., Jayakumar, H., Raghunatah, V.: Quality configurable approximate dram. IEEE Trans. Comput. **66**, 1172–1187 (2017)
19. Jung, M., Mathew, D.M., Weis, C., Wehn, N.: Invited - approximate computing with partially unreliable dynamic random access memory - approximate dram. In: Design Automation of Conference, pp. 100:1–100:4 (2016)
20. Bhalodia, V.: Scale dram subsystem power analysis. Massachusetts Institute of Technology (2005)
21. Venkatesan, R.K., Herr, S., Rotenberg, E.: Retention-aware placement in dram (rapid): software methods for quasi-non-volatile dram. In: International Conference on High-Performance Computer Architecture HPCA, pp. 157–167 (2006)
22. Waterman, A., Lee, Y., Patterson, D.A., Asanovic, K.: The RISC-V instruction set manual, volume i: base user-level ISA. EECS Department, UC Berkeley, Tech. Rep. UCB/EECS-2011-62 (2011)
23. Johnson, R.P.: Contrast based edge detection. Pattern Recogn. **23**(3–4), 311–318 (1990)
24. Chandrasekharan, A., Soeken, M., Große, D., Drechsler, R.: Precise error determination of approximated components in sequential circuits with model checking. In: Design Automation Conference, pp. 129:1–129:6 (2016)

25. Chandrasekharan, A., Soeken, M., Große, D., Drechsler, R.: Approximation-aware rewriting of AIGs for error tolerant applications. In: International Conference on Computer-Aided Design, pp. 83:1–83:8 (2016)
26. Froehlich, S., Große, D., Drechsler, R.: Approximate hardware generation using symbolic computer algebra employing Gröbner basis. In: Design, Automation and Test in Europe, pp. 889–892 (2018)
27. Froehlich, S., Große, D., Drechsler, R.: Towards reversed approximate hardware design. In: EUROMICRO Symposium on Digital System Design (2018)

Chapter 5
Information System for Storage, Management, and Usage for Embodied Intelligent Systems

Daniel Beßler, Asil Kaan Bozcuoğlu, and Michael Beetz

Abstract Embodied intelligent agents that are equipped with sensors and actuators have unique characteristics and requirements regarding the storage, management, and usage of information. The goal is to perform intentional activities, within the perception-action loop of the agent, based on the information acquired from its senses, background knowledge, naive physics knowledge, etc. The challenge is to integrate many different types of information required for competent and intelligent decision-making into a coherent information system. In this chapter, we will describe a conceptual framework in which such information system can be represented and talked about. We will provide an overview about the different types of information an intelligent robot needs to adaptively and dexterously perform everyday activities. In our framework, every time a robot performs an activity, it creates an episodic memory. It can also acquire experiences from mental simulations, learn from these real and simulated experiences, and share them with other robots through dedicated knowledge web services.

The authors are with the Collaborative Research Centre "Everyday Activities Science and Engineering" (EASE), University of Bremen, Bremen, Germany.

D. Beßler (✉)
Collaborative Research Centre "Everyday Activities Science and Engineering" (EASE), University of Bremen, Bremen, Germany

Institute for Artificial Intelligence, University of Bremen, Bremen, Germany
e-mail: danielb@cs.uni-bremen.de

A. K. Bozcuoğlu · M. Beetz
Collaborative Research Centre "Everyday Activities Science and Engineering" (EASE), University of Bremen, Bremen, Germany

C. S. Große, R. Drechsler (eds.), *Information Storage*,
https://doi.org/10.1007/978-3-030-19262-4_5

5.1 Information Systems for Embodied Intelligent Agents

Technology has transitioned to an era where agents are equipped with sensors, actuators, and constant network access. Prominent examples for systems in this era are smartphones, Google glasses, Oculus Rift, and robotic agents. With respect to information acquisition and use, these systems come up with unique characteristics and requirements.

One aspect is that these systems are integrated into perception-action loops where agents execute intentional activities. For example, a robotic agent fetching milk from a fridge has to control its motions based on the image stream captured by its camera. Similarly a Google glasses device accompanying a person shopping in a super market would take the image stream of the camera and visually annotate objects detected in the image with semantic information.

Secondly, these systems are unique in a sense that the information received is always structured according to intentional activities and observed by the acting agent which makes the information highly compressible and generalizable.

In this chapter, we will introduce a conceptual framework in which such information system can be represented and talked about. We will further discuss ongoing research and its impact on information storage management and usage.

5.2 Mastering Everyday Activities

Since the late 1990s and early 2000s, robotic platforms that are physically capable of humanlike manipulations such as Honda ASIMO [32] and Stanford PR1 [40] have started to appear in the market. In the early phases, we have witnessed these robots do complex manipulations with purely teleoperation [40] or a limited autonomy [28]. Then, the research has been shifted toward intelligent robotic applications that, autonomously, carry human scale manipulations, such as chemical experiments [23], hospital assistance [30], and pancake baking [5]. Even though such intelligent applications are big steps toward companion household robots, this alone is not enough for employing robots in our daily lives due to limitations in terms of adaptability, speed, and flawless executions in today's robot control programs. Thus, the next milestone in this area is to enable robots to *master* human-level activities (Fig. 5.1). By using the term *mastering*, we mean being able to carry these manipulations adaptively, dexterously in different environments without having any human assistance or input.

We believe that information processing and advanced cognitive capabilities are keys for achieving this milestone. By combining different required information sources in a coherent information system, the robot is able to reason about past experiences up to the detail of fine-grained motions, to learn abstract knowledge from the situated experiences, and to adapt the abstract knowledge to new situations. On the other hand, efficiently harmonizing and using different information sources with state-of-the-art planning, reasoning, and learning techniques is still a hard science and engineering aspect that needs to be addressed.

Fig. 5.1 Toward robots mastering everyday tasks from performing them

5.3 Information Usage in Everyday Activities

Everyday activities such as household chores are demanding in terms of manipulation skills and dexterity. For instance, preparing a pizza contains subactivities such as dough-rolling and pouring tomato sauce over the dough. Each of these subactivities has its own requirements, which may also vary with respect to environmental properties such as that dough-rolling should continue until the shape of the dough has reached a desired state or that tomato sauce must not spill to elsewhere. To perform such activities under varying conditions, humans use advanced cognitive abilities such as reasoning, recalling, and learning to adjust themselves to new conditions and environmental properties [2].

In order to develop an information system for artificial embodied agents, the human memory system is a blueprint for effective information and knowledge acquisition, storage, and use. On one hand, the human memory system can acquire and generalize knowledge from very few examples [1] and also across different modalities such as watching videos [3] or reading instructions [25]. The second aspect is that humans heavily use episodic memories, which are autobiographic recordings of activities. The human memory system further abstracts and consolidates knowledge through dreaming [39]. Another impressive aspect of the human memory system is the persuasive presence and use of commonsense and naive physics knowledge [17]. For example, just telling a person to set the table is enough for having a detailed intended arrangement of utensils on the table.

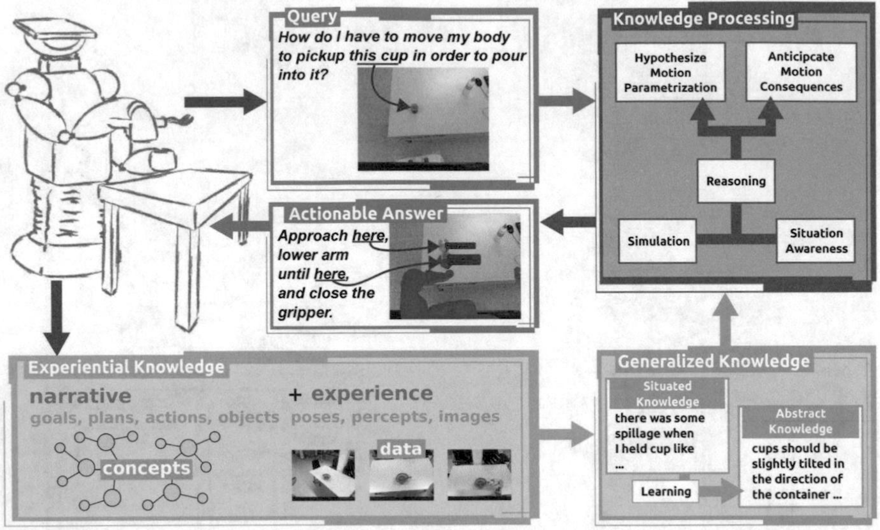

Fig. 5.2 An information system for embodied intelligent systems inspired by capabilities of the human memory system. Knowledge is acquired through experimentation and simulation, generalized through learning, and combined with other information sources to generate answers that are executable by an embodied agent

Having artificial information storage, acquisition management and service with capabilities similar to those of the human memory system would greatly advance the capabilities of technical cognitive systems. Such an artificial information system for embodied cognitive systems is outlined in Fig. 5.2. One of its main components is the question answering interface through which robotic agents can obtain *actionable* answers. With actionable we mean that the answers are specific to the task of the robot, its embodiment, the environment in which the task is performed, and the beliefs the robot has about its world. In the shown example, the main task of the information system is to bridge the gap between the vague and shallow instruction to *pick up a cup* and the situated realization of this activity. The robot has to know how the cup should be approached, where to grasp it, how much force to apply, and how it should be held in order to successfully achieve its goals without causing unwanted side effects. Every time the robot performs such an activity, it can memorize *which* actions were performed and *why* they were performed, and it can remember *how* the actions were performed in terms of body motions, images captured, objects recognized, and whether the goals were achieved or not. This experiential knowledge is highly specific to the situation during which the knowledge was acquired and needs to be generalized to make it applicable in new situations. Finally, the actionable answer is generated based on a combination of different knowledge sources and reasoning mechanisms such as naive physics reasoning, simulation-based reasoning, and learned abstract knowledge.

5.4 An Information System for Embodied Agents

We aim to develop embodied intelligent agents capable of doing complex manipulations in unstructured environments with long-term autonomy. To achieve this, we employ a knowledge-extensive methodology which harmonizes knowledge from different sources such as human activity data, virtual reality, and previous execution logs of robots to provide agents necessary competence to learn, do, and master such complex manipulations in different environments.

Knowledge representation and reasoning (KR&R) in robotic research has the ultimate goal to equip robots with the capability to answer questions such as *"can I perform this action?"*, *"what would be the outcome?"*, and *"how do I have to move my body in order to successfully perform this action?"*. KR&R systems in classical AI would not be able to answer these questions in full detail. Let us consider the blocks world, a classical AI planning example, which is depicted in the left part of Fig. 5.3. Classical AI would approach this scenario as dynamical system, inferring solutions solely based on this homogeneous model. In a logical knowledge base, for example, Boolean queries are true if they are logical consequences of the respective axiomatization in the knowledge base. System dynamics would be axiomatized in terms of conditions under which actions can be performed and which effects these actions might have (see, e.g., [31]). Some other notable homogeneous models are

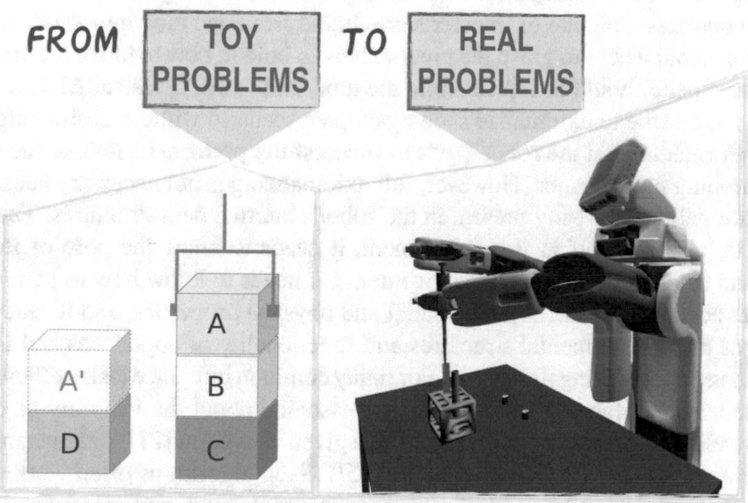

Fig. 5.3 Left: The famous "blocks world" which is traditionally approached using purely logical formalisms and reasoning about the axiomatized scene representation to infer whether all preconditions are met to, for example, move block *A* to position *A'*. Right: Scene visualization of a robot acting in a natural laboratory environment operating a pipette; this domain requires far more complex reasoning involving geometrical and physical features of the objects acted on. However, efficient algorithms exist to compute many task-relevant relations such as reachability or stability

STRIPS- or PDDL-based [15, 24] systems and systems using probabilistic [19] representations.

These classical approaches are very elegant in uniformly handling different problems, but on the other hand tend to have high and unpredictable computational costs due to exponentially growing number of axioms for more complex problems. One prominent example is the egg-cracking problem [26]. The problem is how the activity to crack an egg can be formally described in a logic theory. The theory must guide the agent in its decisions that transform an intact egg into a cracked one while its yolk being still intact and separated from the white in some bowl. Embodied agents would only be able to employ such a theory if it comes with a notion of space and physics—an extremely difficult task in logical formalisms [34]. The task is to reason about how much force to apply to crack the egg, how to move the arm to separate yolk and white, and so on.

One way to avoid the rapid growth in the complexity is to abstract away from geometrical and physical properties of objects. But it is often impossible to find an appropriate level of abstraction because geometrical and physical properties are valuable to embodied agents for reasoning about action feasibility (e.g., reachability or stability of object placements [13]), how the body should be moved, how much force to apply, and which effects can be expected when performing an action.

A robot acting in a realistic and natural environment has to do significantly more complex reasoning than would be possible with a model similar to the block-stacking model. The right part of Fig. 5.3 depicts a more complex scenario in which a robot operates a pipette to transfer some liquid from one tube into another. It has to reason about where to grasp the pipette, how to hold it, how to move the arm, how deep the pipette should be injected into the tube, and so on. Classical AI approaches would not be able to axiomatize such a complex scenario without abstracting away from information that the robot needs to successfully perform its task or sacrificing the reasoning performance. However, full axiomatizing is not necessary because all this information is already present in the robot's internal data structures: The robot needs to localize itself in its environment, it needs to know the pose of its body parts and actuators to coordinate movements, it needs to know how to parametrize motions according to the object's spatial and physical properties, and it needs to be aware of the environmental structures and functionality of appliances and tools in order to use them during its activity. For many common inference tasks, efficient and well-understood algorithms exist such as reasoning about the visibility of objects (i.e., by virtually rendering the scene from a given viewpoint [27]) or reachability of locations (i.e., using inverse kinematics [35]). Such algorithms often work on raw data instead of relying on symbolic abstraction and thus can generate more detailed answers than would be possible if the data is abstracted a priori.

Instead of trying to fully axiomatize, robots shall reuse existing data and algorithms whenever possible and only compute a "symbolic view" of the data on demand given a reasoning task at hand. By doing so, robots can find a situational suitable level of data abstraction while avoiding the problem of finding an appropriate level a priori. Logical inference is then performed on a rather abstract level using the (rather shallow) axiomatization of the robot's world and knowledge about how ontological symbols can be computed on demand by using existing procedures. Thus, the symbolic knowledge base can be seen as an integration layer of heterogeneous reasoning resources whose inference results can be symbolically abstracted to be hooked into the symbolic reasoning. Its logical axiomatization does not need to be as comprehensive as in common AI reasoning methods since a vast amount of information is represented as raw data and linked to the symbolic knowledge base via existing and well-understood procedures.

As a summary, we aim for going beyond the planning and reasoning approaches in classical AI by leveraging knowledge from heterogeneous sources and abstracting it on demand into a coherent symbolic representation. By acquiring information from different sources and abstracting it on demand, we believe that embodied agents can better cope with the complexities and uncertainties in the real world.

In the rest of this section, we first discuss the representational foundation of our approach in Sect. 5.4.1 and how agents can use this information through a uniform logics-based interface language in Sect. 5.4.2. In Sect. 5.4.3, we discuss episodic memories of everyday activities, what information they contain, and how they can be automatically acquired from embodied agents performing everyday tasks. Intelligent decision-making also demands a notion of body, its parts, and capabilities they enable (Sect. 5.4.4) and a notion of environment, tools, and their functionality (Sect. 5.4.5). Commonly used learning algorithms can be employed in order to generalize the information contained in our system and to find implicit information hidden in it (Sect. 5.4.6). Information about embodied systems and their environment may also be used to configure simulation and rendering methods that enable the system to predict the physical state of the environment, and the appearance of it (Sect. 5.4.7). Finally, we explain how agents may share information through a web service for information storage and management (Sect. 5.4.8).

5.4.1 *Structure of Everyday Activities*[1]

Action representations used in AI and autonomous agent research typically represent the agent control system at a coarse level of detail, at which actions are described by black box models. Instead, we put ontologies at the core of the robot control system so that data structures, even those used at lower levels, can be semantically annotated with their meaning according to the core ontologies.

Ontology formalisms are widely used for the representation of *abstract knowledge*, which can be used to describe concepts of the robot's world and which relations might hold between them. A robot working as assistant in a chemical laboratory, for example, may receive an instruction to *neutralize the alkaline substance in some tube*. First, the robot needs abstract knowledge about the phrases "neutralize," "alkaline," and "tube." It needs to know that a "tube" is a container for

[1]Parts of this section were previously published by Beetz et al. [8].

liquid substances with an opening to pour substances into it, "alkaline" is a chemical substance and the counter part of acid, and "neutralization" is a chemical process caused by mixing acid and alkaline substances.

The robot further needs to infer that an acid substance should be used for the neutralization and that a pipette should be used to transport small quantities of acid into the tube. It also has to infer that pipetting is a motion that involves ingesting some acid from a source container into the pipette and dripping it into the tube by pressing a "button" on the pipette. This type of abstract knowledge is often accounted for in ontology representations.

Experiential knowledge, on the other hand, is very detailed and specific information that does neither abstract away from the embodiment of the robot nor from situational context and from which general knowledge can be derived. This is, for example, that the tip of the pipette should be inserted into the tube or at least held close above it to avoid spillage. This type of knowledge can be obtained through practical experimentation and mental simulation. It is substantial information for finding appropriate motion parameters that are likely to achieve desired effects. Representations for integrated robot control systems should therefore take into account both abstract and experiential knowledge.

We consider the experiential knowledge required for intelligent behavior realization as component of the ontological characterization of the robot's experiences. Robots shall build up a knowledge base that gains experience whenever the robot performs an activity and gather information about previous experiences, generalize over specific situated experiences, and adapt them into the situated context ahead. We argue that a linked ontological specification of the distinct kinds of knowledge required is a powerful, elegant, and highly integrative framework for robotic agent question answering.

In order to gain a better intuition of the advantages of putting symbolic ontologies at the core of robot control systems, consider, for example, the concept of a dynamically changing robot pose. We represent this in the form of the *holds* predicate $holds(pose(r,\langle x,y,o\rangle), ti)$, which asserts that the robot believes its pose with respect to the origin of the environment map at time instant ti is $\langle x,y,o\rangle$. Typically, the robot control system would estimate the pose of the robot over time using a Bayesian filter—such as a particle filter for robot self-localization. In this case, the robot's belief about where it is implemented in the probability distribution over the possible robot poses in the environment is estimated through the filter. Then, one can specify the pose where the robot believes to be ($holds(pose(r,\langle x,y,o\rangle), ti)$) as the pose ⟨x,y,o⟩ with the maximal probability. By specifying rules that ground concepts of the ontology into the data structures of the control systems, substantial parts of the data structures can be turned into a virtual knowledge base where the relevant relations are computed from the data structures on demand.

The core ontology of our robot control system defines (•) robots, their body parts, how the parts are connected, and sensing and action capabilities; (•) objects, their parts and functionality, and constellations and configurations thereof; (•) robot tasks, actions, activities; and behaviors, and (•) situational context and environment. The ontology together with additional axioms and rules provides background knowledge that is relevant for manipulation tasks. For example, it states that a cup is a vessel that consists of a hollow cylinder and a handle and that it can be used for drinking, mixing, and pouring substances. It also provides knowledge about the material they are made of, namely, metal, wood, porcelain, or plastic. A key role of the ontology is also the grounding of the ontology concepts in different components of the control system: the perception, reasoning, and control components.

One of the big advantages of using ontologies is that different ontology modules are easily combinable (as long as they are based on the same foundational ontology). Additional special-purpose ontologies for robots can be used to gain more application domain information. Outdoor robots, for example, can use ontologies developed for geo-information systems, such as the ontologies of *OpenStreetMap* tags. Robotic assistants for department stores can use product data from web stores such as *GermanDeli*,[2] etc.

5.4.2 *An Interface Layer for Embodied Agents*[3]

Cognition-enabled robotic agents can be viewed as robot control systems that *know what they are doing* [12]. This can be achieved by representing models of the robot, the control system it employs, the task, objects, and environment as well as the interactions between them in a knowledge representation language. The inference mechanisms of the knowledge system can then be employed to automatically answer queries regarding to what the robot does, how it does it, why it does it, what might happen, what are alternative ways of performing an action, and so on [6].

Much of the information used in robot control systems can be represented, to some extent, at the abstract or experiential level in the knowledge base. Robots, unlike many other cognition-enabled systems, have further access to internal information acquired through processes of their control system. For example, the robot might have the abstract belief to be at some location but at the same time, this information is also implicitly contained in the probabilistic estimation used by the navigation system. This raises two important issues. First, how can we ensure that the abstract beliefs of the robot are consistent with the information contained in its control system? Second, how can we ensure that if a robot performs an action that requires certain knowledge that this knowledge is also present at the abstract level?

One way of dealing with these two problems is the notion of virtual knowledge bases. Virtual knowledge bases are symbolic knowledge bases that can compute abstract representations on demand from information contained in the robot's control system or by running computation processes. The realization of virtual

[2] http://www.germandeli.com.

[3] Parts of this section were previously published by Beetz et al. [8].

knowledge bases in robot control systems is complicated as information is held in different system components and in different forms and data structures. This raises the issue whether we can ensure a consistent operation with these different implementations of the same information and hide the complexity from the programmers of robot applications.

This problem was encountered previously in different areas of computer science. The introduction of an *interface layer* has in many cases proved to be a promising approach to dealing with these issues. The purpose of interface layers is to provide a stable standardized and concise abstraction interface that decouples the development of applications from further developments in the technological basis. One of the most prominent examples is the relational database concept as an interface layer between database implementation and advanced information technologies.

The mastery of everyday manipulation activities requires robots to be equipped with sufficient knowledge and the ability to draw conclusions based on very heterogeneous mechanisms [4]. The accomplishment of more complex reasoning tasks often requires the combination of information from different inference mechanisms into a coherent picture or to make a decision in the face of contradictory information. This complexity is difficult to manage by robot application programmers who often rely on third-party mechanisms and the (often not standardized) representations they employ.

To this end, we employ a uniform logic interface to the heterogeneous representations and reasoning mechanisms of robot control systems which is depicted in Fig. 5.4. This interface language presents the hybrid reasoning kernel to the programmer as if it were a purely symbolic knowledge base. It takes an object-oriented view, in which everything is represented as if it were entities retrievable by providing partial descriptions for them. Entities that can be described include objects, their parts, and articulation models, environments composed of objects, software components, actions, and events.

The interface layer facilitates the use of heterogeneous representations and reasoning mechanisms through a uniform logic language. It appears to robot application programmers as a conventional first order logic (FOL) -based language, but symbols in queries are potentially computed on demand and abstracted from representations used by procedures of the robot control system. An example usage of the interface layer for question answering is depicted in Fig. 5.5. The query corresponds to a question such as *"at the end of which pipetting actions was the tip of the pipette inside of the tube into which the substance was pipetted?"*. It expands to qualitative spatial reasoning about the pose relation symbol that is abstracted on demand from memorized data structures of the control system. Many relevant inferences require to combine information pieces of different types and from different sources. Before grasping a pipette, for example, the robot may ask *"am I able to grasp the pipette from my current position?"*. This reasoning task involves to infer possible grasping points using the perceived object pose, to read the position sensors of some joints, and to infer the ability to grasp using inverse kinematics. These heterogeneous information pieces are made available through the logic-based interface language and thereby can be combined on the symbolic level

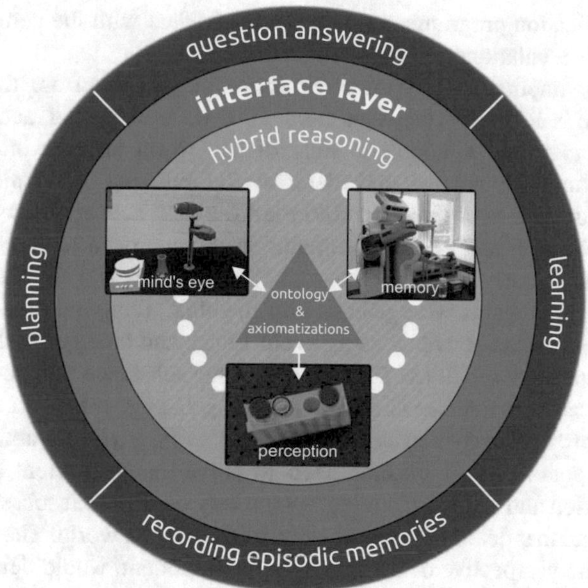

Fig. 5.4 An interface layer for embodied agents that exploits logical representations at the core of robot control systems. All mechanisms integrated below the layer automatically become available to the applications above it

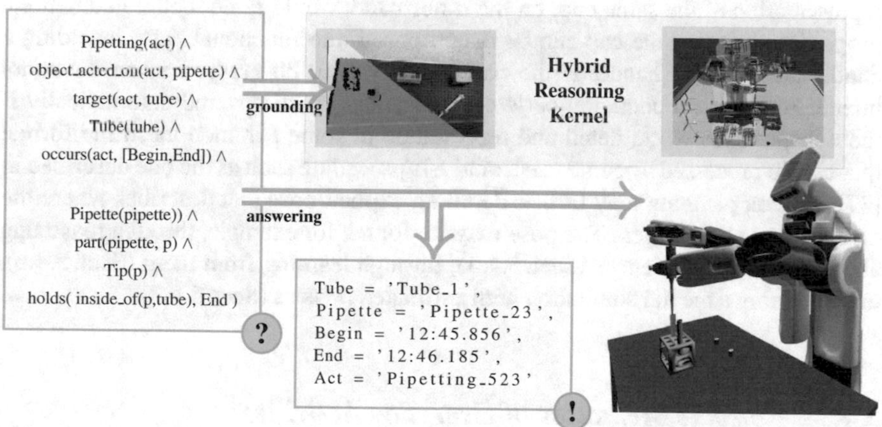

Fig. 5.5 Symbolic queries are logical conjunctions that are grounded in heterogeneous mechanisms of the robot's control system and through logical reasoning. The answer is given as set of bindings for free variables appearing in the query and in the form of visualizations

such that application programmers do not have to deal with the complexity of the underlying representations and mechanisms.

Particularly important for competent robot manipulation are the events that occur and the state properties that hold at certain stages of action execution (predicates *occurs* and *holds*). The state of the world and the internal state of the robot continuously change, and therefore the outcome of all rules attached to this information. The robot may need to plan the next action, for example, during clearing a table, and infers that the remains in the cup need to be poured into the sink, that it has to carry the cup to the sink, and that it has to tilt it and wait a few seconds until the remains were poured into the sink. Temporal predicates express facts about occurrence of events using event logics and changes in the world state using time-dependent relations. The fact that some substance was poured at 11 am, for example, can be written as $occurs(pouring_0, 11\text{ am})$ or the fact that the cup is empty afterward can be written as $holds(empty(cup_0, true), 11\text{ am})$.

Integrated robot control systems need to solve many different tasks, such as visual perception and motion planning, which rely on different subsets of available information and that detail different aspects of the robot's world. The description of a cup, from the perspective of the perception component, would detail a subset of the object's visual properties such as its shape or primary color. These properties can be computed by specialized object recognition procedures that operate on the raw stream of visual sensor data in order to find a symbolic abstraction. The primary color, for example, could be estimated by computing a color histogram and picking the dominant color as the primary color of the object. The controller-centric representation of the same cup, on the other hand, would rather detail geometrical properties such that the cup can be decomposed into functional parts including a handle and that the handle is the part that needs to be grasped when the robot intends to hold it. Functional parts of objects can be discovered automatically if their shape is known in detail and represented in some common mesh file format that can be processed by some mesh analysis procedure such as the one described in [37]. Motion planning may further detail an explicit pose that describes where the object should be grasped. The pose may be found, for example, through reasoning about previous experiences (Sect. 5.4.3), through learning from them (Sect. 5.4.6), or by running a mental simulation with estimated physics (Sect. 5.4.7).

5.4.3 Episodic Memories of Everyday Activities[4]

When somebody talks about the final goal in the last soccer world championship, many of us can "replay" the episode in our "mind's eye." The memory mechanism that allows us to recall these very detailed pieces of information from abstract descriptions is our episodic memory. Episodic memory is powerful because it allows

[4]Parts of this section were previously published by Beetz et al. [8].

us to remember special experiences we had. It can also serve as a "repository" from which we learn general knowledge.

Episodic memories shall be deeply integrated into the knowledge acquisition, representation, and processing system. Whenever a robotic agent performs, observes, prospects, and reads about an activity, it creates an episodic memory. An episodic memory is best understood as a video that the agent makes of the ongoing activity coupled with a very detailed story about the actions, motions, their purposes, effects, the behavior they generate, the images that are captured, etc.

We call the episodic memories created by our system narrative-enabled episodic memories (*NEEM*s). A *NEEM* consists of the *NEEM experience* and the *NEEM narrative*. The *NEEM experience* is a detailed, low-level, time-indexed recording of a certain episode. The experience contains records of poses, percepts, control signals, etc. These can be used to replay an episode in detail. *NEEM experiences* are linked to *NEEM narratives*, which are stories that provide more abstract, symbolic descriptions of what is happening in an episode. These narratives contain information regarding the tasks, the context, intended goals, observed effects, etc.

An example of the information contained in a *NEEM* is illustrated in Fig. 5.6. In this episode, a robot cleared a dinner table that had a bowl and a spoon on top. The depicted timeline has marks for some time instants at which the robot started to perform an action. Images that were captured by the robot at these time instants are shown on the top row of the figure. The poses of the robot are shown below. The robot navigates to the table at t_1 to perceive objects on top of it at t_2, to establish a pregrasp pose at t_3, and to grasp the spoon at t_4. Some of the corresponding assertions in the knowledge base are shown at the right side of the figure. These assertions represent, for example, that at t_2 an event ev_{123} occurred, that this event was a detection event with the corresponding perception task obj_{246}, that the perceived object is described by obj_{345}, and that this object corresponds to the image region reg_{567} of the captured image img_{456}. The symbolic assertions are linked to data structures of the control program to enrich the high level activity description with low-level information such as concrete motions.

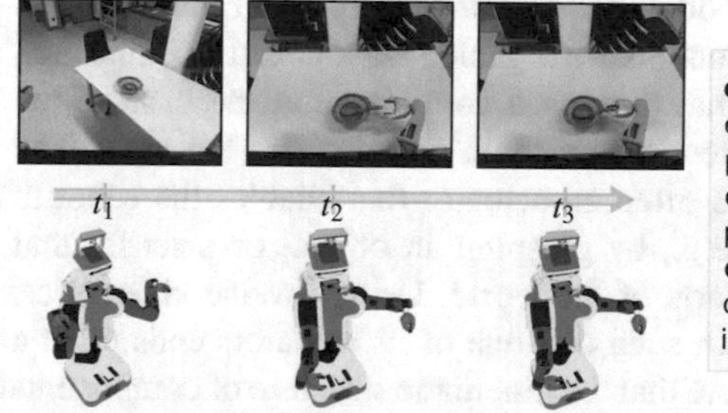


```
occurs(ev123,t2)
event-type(ev123,detect)
perception-task(ev123,obj246)
entity(obj246, [an, object, ...])
perception-result(ev123,obj345)
entity(obj345, [an, object, ...])
captured-image(ev123, img456)
image-region(obj345, reg567)
```

Assertions

Fig. 5.6 Illustration of a narrative-enabled episodic memory (NEEM). Left: The NEEM experience in the form of a time series of control-level data. Right: The NEEM narrative describing the activity on a higher level with symbolic assertions in the knowledge base

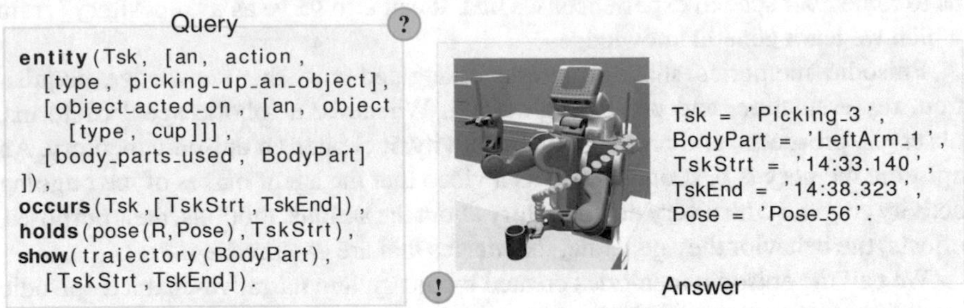

Fig. 5.7 Example query evaluated on a NEEM on the left and the respective answer on the right

*NEEM*s allow to ask queries about which actions the robot performed; when, how, and why they were performed; whether they were successful; what the robot saw; and what the robot believed when the action was performed. The robot may ask queries such as: *"how did I pick up a cup?"*, *"which body part did I use?"*, and *"how was my pose when picking it up?"*. These questions map to a query such as the one depicted in Fig. 5.7. Here it searches for *NEEMs* where Tsk is a task during which the robot picked up a cup with its body part $BodyPart$, that occurred during the time interval $[TskStrt, TskEnd]$, and at which start time the pose of the robot is described by $Pose$. The answer to that query can be given in terms of symbol bindings for the free variables in the query and visually by rendering the scene based on beliefs of the robot.

5.4.4 Body Representation in Everyday Activities

Self-models are substantial for intelligent agents to experience the world through a first-person perspective. The body of an agent consists of parts such as the elbow, wrist, and so on that are grouped into more complex components such as the hand, arm, and upper body. Different body models are depicted in Fig. 5.8. From a motion controller perspective, hardware components are organized in kinematic chains (i.e., through some form of joints), such that moving a component at the base of the chain (e.g., shoulder) will move the rest of the chain with it (e.g., the arm). The last component in kinematic chains is often an actuator that enables the robot to manipulate the world in some way (e.g., by grasping an object) or a sensor that enables the robot to perceive some parts of the world. Using inverse kinematics, the robot may find a collision-free path such that one of its actuators ends up at a desired target pose (motor control). Note that the kinematic structure of components is inherently required information for successful motor control. Motor control plays a central role for cognition in mobile robotics and in particular in the perception-action loop (i.e., sensorimotor integration). Action awareness and information about the own body are crucial for generating complex, flexible, and adaptive behavior that

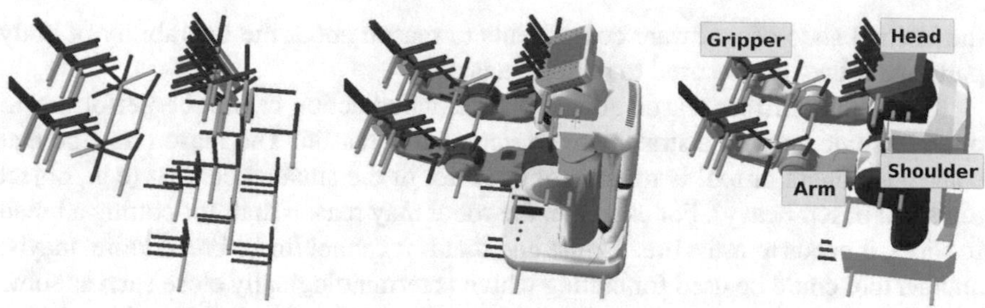

Fig. 5.8 Different perspectives on embodiment. The body of the robot is represented (from left to right) as (•) kinematic chain, (•) collection of meshes, poses, and visual attributes; and (•) semantic component descriptions

controls the body as a whole instead of operating actuators as if they were attached to some static holder without any interrelation. Action-aware motor control is a step toward motor cognition: While motor control can answer the question of *how* the robot can operate its motors to reach a specific pose, motor cognition can answer the question of *why* the robot wants to reach that pose and to find better options given the situational context which would be hard to account for in pure motor control frameworks.

The embodiment of robots is heavily influenced by their purpose and how the bodies can be used to perform some (often specific) tasks. Be it an entertainment-centered platform with a curvy body and facial expressions but weak actuators that rarely can perform any manipulation task or an industrial robot with powerful arms that is tailored to lift heavy car pieces very precisely during assembly in a restricted assembly hall. This is contrary to the rather homogeneous configuration of human bodies which are similar in size and shape and also in terms of capabilities. Self-models are in particular important for robots acting in less restricted environments such as regular kitchens. In kitchens, robots can perform best when they mimic the human body: The body size and arm length influence which places are reachable; the shape, precision, and strength of the gripper which objects are operable; and so on.

Adaptive robots need to be informed about their body, which components are available, and how these can be used to perform a situated action to answer questions such as *"am I able to do this?"*, or *"what can I do with my body?"*. To answer these questions, the robot needs to be informed about the relation between its components and its capabilities to perform certain actions. This is, for example, that the robot needs wheels or legs for locomotion or an end effector for grasping. Components can be further restricted in terms of which actions can be performed with them under which circumstances. (e.g., the component must be turned on and nonbroken, no human shall be harmed, etc.) This is to implement robust robot behavior even in the face of unforeseeable circumstances such as broken hardware, and that the system stays operable in case it has some redundancy that compensates the damage or at least is being able to conclude its inoperability. To this end, robots may monitor the internal state of hardware components to reason about the operability of body parts and if they can be used to perform a task.

Robots may further reason about *why* a situated action cannot be performed in order to come up with a strategy to overcome this reason. The reason may be that some component or tool is missing or does not fit the situated context (e.g., object too small or too heavy). For example, the robot may reason that, for cutting a bread in slices, it needs to use a bread knife and that if it cannot find a bread knife, maybe another tool could be used for cutting which is terminologically close such as some other types of knife or, more general, objects with a blade and a handle to be grasped. Robots may also be able to (semipermanently) install hardware components or tools if they have extendible actuators. The functionality of installed components and tools is inherited by the robot such that its capabilities to perform certain actions are extended.

Another valuable information that can be considered as part of the robot's self-model is the software that it can use to process the sensor data, control its body, plan its activities, etc. A camera device alone, for example, does not solely make the robot recognize objects in the world and relate them to its activity. The camera only provides high-volume, continuous space image streams that need to be segmented and further analyzed to recognize occurring objects in the scene. Thus, a camera alone does not give the robot the capability to perceive objects; an additional software component is required that implements the abstraction of continuous space data into ontological symbols in the knowledge base (e.g., the class labels of perceived objects). Software use is often restricted to specific hardware configurations. In the case of visual perception, the perception framework may only work with specific camera properties (e.g., minimum or maximum image resolution) or under specific environmental circumstances (e.g., in case some model was trained that is not environment-invariant). Robots could further employ automated and situated discovery and installation of software components to gather new capabilities. Many software repositories exist, but no standard for the meta-information required for situated software discovery for robots has been established so far.

5.4.5 Environment Representation in Everyday Activities

Robots that do complex manipulations in unstructured environments highly rely on environmental information. First of all, robots need to be able to answer the questions *"where am I?"* and *"where should I be to perform this action?"*. To answer these questions, robots need to localize themselves within their environment and reason about how locations relate to their activity. For instance, if a robot needs to fetch a milk package from a fridge, it first needs to find its own pose relative to the fridge, then it needs to navigate toward a place nearby the fridge so that it can open the door of the fridge. In addition, these maps can be used to predict poses of objects which cannot be perceived by the perception systems due to being invisible from the perspective of the robot.

Environmental and object features highly affect robot plans and motion parameters. Robots need information about those features in order to find an appropriate parametrization of their control program that is promising for achieving some designated goal. In case of fetching a milk package from a fridge, for example, the robot needs to know geometrical and physical properties of the fridge such as the shape of the door handle and its location in order to parametrize its opening motion.

In our framework, such environmental knowledge is represented as 3D maps together with semantic information about the environment. The semantic information contains relevant properties of objects located in the environment. Foremost, each object in the environment is linked to a concept in a taxonomy. At the more general levels of this taxonomy, we have concepts such as *Physical Device* (i.e., a rigid artifact with a specific function) or *Food* (i.e., objects eaten by some life-forms). At these levels, we can also define the concepts in more detail, for example, that the primary function of food is to be ingested or that electrical devices need a power supply, etc. Definitions at the more general levels of the taxonomy are inherited by the more specific concepts (subsumption hierarchy). One example is depicted in Fig. 5.9. In this example, a particular fridge perceived by the robot is linked to the concept *Refrigerator* which is defined in the ontology. The concept further defines that refrigerators are the primary storage place for objects of type *Perishable* and that *Milk* is a perishable drink. The *Milk* concept may again be linked to a particular perceived milk package, for example, when the robot needs to find a storage place for items remaining on the table after a breakfast event.

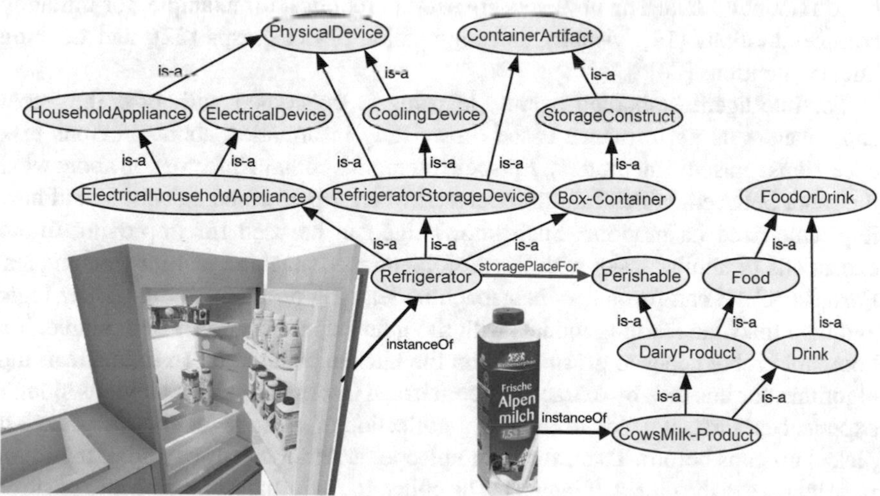

Fig. 5.9 An illustration of semantic map entities corresponding to perceived objects and how they are linked to concepts in a taxonomy

Semantic maps employ very detailed part and shape descriptions of the objects which include articulation information for the objects consisting of movable parts such as drawers and cupboards. Articulation information between a part and its respective root object is explicated through *joint* relations. For instance, there might be a *hinged joint* between a fridge and its door. Objects may also be linked to 3D mesh data. Using these geometrical models, it is possible to infer part decomposition and articulation information of objects [37]. Meshes are also useful blueprints for the perception system to recognize the objects.

5.4.6 Learning from Episodic Memories[5]

Machine learning is a field in computer science whose purpose it is to create programming routines that accomplish given goals using statistical methods instead of being explicitly programmed to do that. Nowadays, it has being widely used in many areas where programming explicit algorithms with a desired performance level is not feasible, such as spam email detection and computer vision. Most, if not all, of the machine learning techniques are data-extensive approaches that require a set of previous samples in order to produce statistical models of the given problems.

In robotics, machine learning is heavily used in order to "teach" robots how to interact with the world. Learning applications in robotics include teaching an autonomous helicopter how to fly [21], making a robotic arm grasp novel objects [33], and teaching a biped robot to walk [9]. Recently, a new set of techniques, called deep neural networks (DNN), is popular in the field of machine learning. These techniques train artificial neural networks to identify patterns in huge amount of data. The networks are used in robotics, for example, for imitating complex motions [14], visually detecting stable object grasps [22], and tracking human intentions [20].

For intelligent embodied agents, learning is concerned with how the agent can enhance its performance based on existing information about previous task executions (episodic memories). Episodic memories contain information about what the robot believed, which failures occurred when it performed an action, and how it parametrized its motions. Such knowledge can be used for improving future executions of similar tasks with the aforementioned machine learning techniques. For this, robots can formalize their machine learning problems in first-order logic and train machine learning models with the inferred information. For example, if a household robot needs to grasp a cup on the kitchen counter, it can train a learning algorithm for this task by extracting geometrical information from previous similar experiences. First, it needs to retrieve a collection of episodic memories of how it picked up cups before. Then, for each episode, it retrieves, for example, the robot pose relative to the object of interest. The collection is further separated into positive

[5] Parts of this section were previously published by Bozcuoglu and Beetz [10].

and negative samples according to whether the corresponding task is associated with any failures. As the last step, the desired learning algorithm is trained using this dataset.

Learning mechanisms shall be tightly coupled with knowledge representation and reasoning systems of robotic agents. This is to use the knowledge base to provide information about situational context of episodic memories for learning and to use the reasoning infrastructure for situation-aware information selection. It is, for example, not always the case that the latest generated trajectory is better than the previous ones in an iterative trajectory learning process. By having some trajectories available that were learned from episodic memories, robots can compare, for example, their durations and lengths and then execute the one that fits the most in the current situation.

Another advantage is that information about situational context can be exploited for the adaptation of learned knowledge to new situations, environments, and robotic platforms. For instance, semantic environment maps (see Sect. 5.4.5) are useful for adapting motions to different environments. They are part of the situational context of episodic memories. A robotic agent may compare the geometrical properties of memorized and current environment and reason about how memorized motions can be adapted to the current situation [11]. As an example, let us consider an episodic memory of serving food from a fridge. A robot, located in a different kitchen, can compare the physical properties of memorized and perceived fridge and adapt the memorized motions accordingly. In case the fridge is wider, for example, it increases the radius of the door-opening trajectory, it uses its other arm and a reflected trajectory if the handle is on the other side of the door, and so on.

5.4.7 Inner World Reasoning[6]

We believe that symbolic reasoning is necessary but not sufficient to realize the full range of reasoning capabilities needed for mastering everyday activities. Thus, we propose to complement the symbolic knowledge representation and reasoning system with an additional knowledge system that can perform subsymbolic reasoning tasks including visual imagination, mental simulations of actions, learning from observation, and semantic retrieval of subsymbolic information about objects, substances, actions, motions, and their physical effects.

The key strength of such a knowledge processing system is that it unifies a collection of bleeding edge reasoning mechanisms through symbolic representations at a level of detail which enables embodied agent's control-level reasoning. Robotic agents are supposed to assert their beliefs about the world in a game engine by accessing the engine's world state data structure, annotating relevant data structures with symbolic names, and asserting symbolic facts about these data structures and

[6]Parts of this section were previously published by Haidu et al. [16].

their relations. The reasoning mechanisms can view the world state as a (virtual) symbolic knowledge base, where the mechanisms of the game engine, namely, data structure retrieval, physical simulation, and rendering, are the essential reasoning mechanisms.

The core component of this knowledge processing system, called *inner world*, performs a basic loop with the following steps:

1. Update the agent's dynamic state (e.g., the control signals send to the joint motors of a robotic agent); possibly there is a world process that also changes the state of the world (e.g., other agents not under control).
2. Update the world state based on the current state and the control inputs generated by the agent and the world process (according to the laws of physics implemented by the physics simulation).
3. Visually render the updated world state.

This loop evolves the virtual world state, which is accessible through the application programming interface of the game engine.

The core is enclosed by a state and event abstraction layer. The state and event abstraction layer abstracts the world state into a representation that facilitates naive physics and qualitative reasoning [36]. To do so, the layer automatically computes physical and spatial relations as well as force-dynamic events. An example of a naive physics relation is the *supported-by* relation. An object O is supported by object S if O is physically stable, in contact with, and above S. Other relations are doors and containers being opened or closed, substances being spilled, etc. Touching is an example of a force-dynamic event which happens when the hand makes contact with another object. The detection of force-dynamic events is essential for the recognition and the segmentation of actions into motion phases. For example, grasping and lifting an object is characterized by the hand touching an object, keeping contact with the object, and the object losing the contact to its supporting surface. An example episodic memory is depicted in Fig. 5.10. It shows that experience data is segmented according to monitored force events such as that the hand gets into contact with a milk package.

State and event abstraction is used by the knowledge representation interface layer that provides modules for activity parsing and recording, query answering, semantic environment model extraction, and virtual image capturing. The activity parsing and recording module takes the stream of time-stamped world states together with the abstract world states, force-dynamic events, and motion events from the agent and generates a symbolic activity representation stated in a first-order time interval logic. The symbolic activity representation is time synchronized with subsymbolic stream data that includes the agent's and object's poses and shapes and images. The semantic environment extraction module maps over the data structures of the world state and asserts for each relevant object and its part's symbolic names, the label category, the part of hierarchy, the articulation chains and models, and other relevant symbolic relationships. The virtual image capturing module can place cameras in the game environment; access their scene's built-in rendering information, such as color, depth, specular, image data, etc.; and further extend it

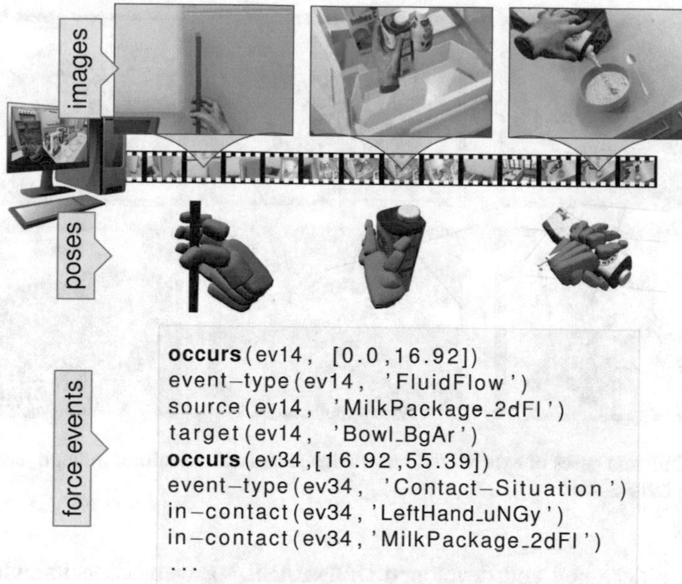

Fig. 5.10 Episodic memory of a human preparing cereals for a breakfast in simulation

by segmenting the image into objects and labeling them with their corresponding symbolic names.

The main advantage of having a "mind's eye" [18, 29] simulation is being able to predict outcomes before execution. Robots that do everyday manipulation tasks can hugely benefit from this prediction. The main requirement for this is being able to operate in a physics-enabled and photorealistic inner world environment and to map the virtual experiences to the real world.

5.4.8 Information Storage and Management Infrastructure[7]

Providing an information system for embodied agents is a difficult and tedious programming task that might require proficiency in some fields of artificial intelligence and data management. Without having a background in these fields, the barriers for employing such a system are very high. Building upon unified representations of robot control systems, such a platform can be designed very generically and be reused for many different robotic platforms, environments, and activities.

[7]Parts of this section were previously published by Beetz et al. [7] and Tenorth et al. [38].

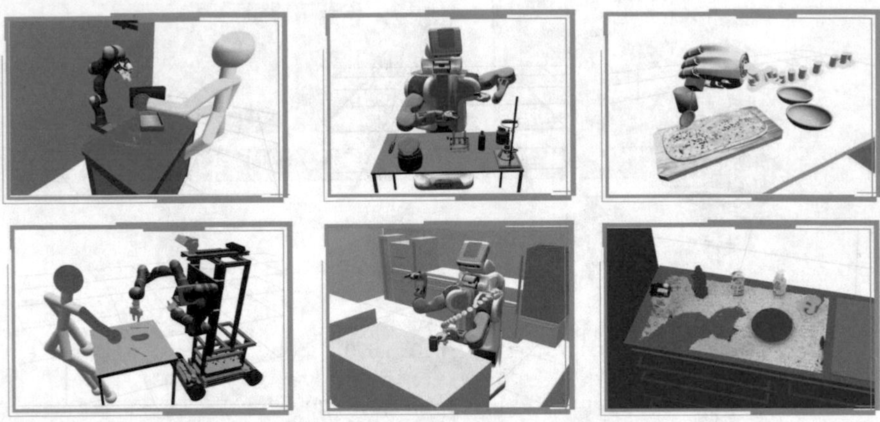

Fig. 5.11 Different cases of experiential knowledge stored and visualized through the knowledge web service OPENEASE

We have designed and developed OPENEASE,[8] a web-based knowledge representation and processing service. It equips autonomous robots with knowledge and reasoning mechanisms contained in our information system. It also provides the infrastructure to make inhomogeneous experience data from robots and human manipulation episodes accessible and is complemented by a suite of software tools that enable researchers and robots to interpret, analyze, visualize, and learn from episodic memories. Users and robots can retrieve episodic memories, ask questions about them, and visualize the respective answers in the canvas of the web front end. Figure 5.11 depicts a collection of episodic memory visualizations generated by the OPENEASE platform. These are episodes of robots performing tasks in an environment with a human coworker, conducting chemical experiments, setting a table in a kitchen environment, and performing perception tasks.

Humans have developed an episodic memory that records experiences in a very comprehensive and detailed fashion in order to provide the informational basis for autonomous learning. Artificial learning so far typically works within a carefully defined mathematical framework without the agent being able to look beyond the structural, representational limits of said framework. Further, the datasets that are used for learning are often too narrow in their scope. If a robot wants to learn flipping pancakes, for example, it is not sufficient to learn from the position, velocity, acceleration, and force data of the motion. The learning data also needs to include the intentions, beliefs, perceived information, predicted effects, etc. Our information system provides an infrastructure for enabling robots to collect episodic memories that do not only provide the pose and sensor data streams but also a symbolic structure on top of this data that enables the robots to reason about what they did, how, what happened when they did it, why they did it, what they believed,

[8]http://openease.org.

and what decisions they made. Robots can also upload information, share it with others, and use information acquired by other robots. We have used this interface, for example, to connect two robots located in different continents [11]. The robot located in Germany has opened a fridge door by inferring how to open it using episodic memories acquired by the robot in Japan.

5.5 Future of Information Systems for Embodied Agents

Advancements in technology inspire us to dream and realize new ideas that seem to be impossible beforehand in order to help humans throughout their life. Especially, technologies that are related to information science are the ones that changed human life in recent years. For instance, gaming technologies and computer graphics have allowed us to represent environments almost photo-realistically, with approximate realistic physics and at low cost. Big data storages allow to save images and other high-volume data and data science being able to generate general knowledge from huge sets of data. Moreover deep learning provides appropriate and efficient memory representations for the respective task.

Within the scope of our work, we make use of similar technologies for generating cognition-enabled robot control which also offer robots the possibility of employing information processing for mastering everyday activities. This means that robotic agents can execute a large spectrum of tasks in different environments by employing the necessary information in their control and planning frameworks and making it actionable in specific and even novel tasks. They do not only retrieve information but also can adapt to the specifics of the situation ahead, even if the specifics were not known before.

We believe that such an information system will help robotic agents to cope with the challenges of everyday activities and transition them from just performing tasks to mastering their jobs. Just like a kid learning how to pour, robots shall acquire information through experimentation and mental simulation. And just like a kid, the robot will do mistakes, memorize them, and try to avoid them in the future. But it will also succeed in specific situations, and it will try to mimic and adapt its successful experiences to novel situations. After many thousand attempts, the robot, just like the kid, will eventually become a master of pouring. And eventually robots will then become capable of assisting humans in their daily life, toward an era in which robots are companions of their human operators.

References

1. Anderson, J.R., Schooler, L.J.: Reflections of the environment in memory. Psychol. Sci. **2**(6), 396–408 (1991)
2. Anderson, J.R., Greeno, J.G., Reder, L.M., Simon, H.A.: Perspectives on learning, thinking, and activity. Educ. Res. **29**(4), 11–13 (2000)

3. Anderson, J.R., Myowa-Yamakoshi, M., Matsuzawa, T.: Contagious yawning in chimpanzees. Proc. R. Soc. Lond. B: Biol. Sci. **271**(Suppl. 6), S468–S470 (2004)
4. Bateman, J., Beetz, M., Beßler, D., Bozcuoglu, A.K., Pomarlan, M.: Heterogeneous ontologies and hybrid reasoning for service robotics: the ease framework. In: Third Iberian Robotics Conference, ROBOT'17, Sevilla (2017)
5. Beetz, M., Klank, U., Kresse, I., Maldonado, A., Mösenlechner, L., Pangercic, D., Rühr, T., Tenorth, M.: Robotic roommates making pancakes. In: 2011 11th IEEE-RAS International Conference on Humanoid Robots (Humanoids), pp. 529–536. IEEE, Piscataway (2011)
6. Beetz, M., Jain, D., Mösenlechner, L., Tenorth, M., Kunze, L., Blodow, N., Pangercic, D.: Cognition-enabled autonomous robot control for the realization of home chore task intelligence. Proc. IEEE **100**(8), 2454–2471 (2012)
7. Beetz, M., Tenorth, M., Winkler, J.: Open-EASE – a knowledge processing service for robots and robotics/AI researchers. In: IEEE International Conference on Robotics and Automation (ICRA), Seattle (2015)
8. Beetz, M., Beßler, D., Haidu, A., Pomarlan, M., Bozcuoglu, A.K., Bartels, G.: Knowrob 2.0 – a 2nd generation knowledge processing framework for cognition-enabled robotic agents. In: International Conference on Robotics and Automation (ICRA), Brisbane (2018)
9. Benbrahim, H., Franklin, J.A.: Biped dynamic walking using reinforcement learning. Robot. Auton. Syst. **22**(3–4), 283–302 (1997)
10. Bozcuoglu, A.K., Beetz, M.: A cloud service for robotic mental simulations. In: International Conference on Robotics and Automation (ICRA), Singapore (2017)
11. Bozcuoglu, A.K., Kazhoyan, G., Furuta, Y., Stelter, S., Beetz, M., Okada, K., Inaba, M.: The exchange of knowledge using cloud robotics. Robot. Autom. Lett. **3**(2), 1072–1079 (2018)
12. Brachman, R.J.: Systems that know what they're doing. IEEE Intell. Syst. **17**(6), 67–71 (2002)
13. Davis, E.: A logical framework for solid object physics. Tech. Rep. 245 (1986)
14. Duan, Y., Andrychowicz, M., Stadie, B., Ho, O.J., Schneider, J., Sutskever, I., Abbeel, P., Zaremba, W.: One-shot imitation learning. In: Advances in Neural Information Processing Systems, pp. 1087–1098 (2017)
15. Fox, M., Long, D.: PDDL2.1: an extension of PDDL for expressing temporal planning domains. J. Artif. Intell. Res. **20**, 61–124 (2003)
16. Haidu, A., Beßler, D., Bozcuoglu, A.K., Beetz, M.: Knowrob-sim – game engine-enabled knowledge processing for cognition-enabled robot control. In: International Conference on Intelligent Robots and Systems (IROS). IEEE, Madrid (2018)
17. Hegarty, M.: Mechanical reasoning by mental simulation. Trends Cogn. Sci. **8**(6), 280–285 (2004)
18. Ishai, A., Haxby, J.V., Ungerleider, L.G.: Visual imagery of famous faces: effects of memory and attention revealed by fMRI. Neuroimage **17**(4), 1729–1741 (2002)
19. Kaelbling, L.P., Littman, M.L., Cassandra, A.R.: Planning and acting in partially observable stochastic domains. Artif. Intell. **101**(1–2), 99–134 (1998)
20. Kim, B.: Interactive and interpretable machine learning models for human machine collaboration. Ph.D. Thesis, Massachusetts Institute of Technology (2015)
21. Kim, H.J., Jordan, M.I., Sastry, S., Ng, A.Y.: Autonomous helicopter flight via reinforcement learning. In: Advances in Neural Information Processing Systems, pp. 799–806 (2004)
22. Lenz, I., Lee, H., Saxena, A.: Deep learning for detecting robotic grasps. Int. J. Robot. Res. **34**(4–5), 705–724 (2015)
23. Lisca, G., Nyga, D., Bálint-Benczédi, F., Langer, H., Beetz, M.: Towards robots conducting chemical experiments. In: 2015 IEEE/RSJ International Conference on Intelligent Robots and Systems (IROS), pp. 5202–5208. IEEE, Piscataway (2015)
24. McDermott, D.: The formal semantics of processes in PDDL. In: Proceedings of the ICAPS Workshop on PDDL (2003)
25. McLaughlin, B.: "Intentional" and "incidental" learning in human subjects: the role of instructions to learn and motivation. Psychol. Bull. **63**(5), 359 (1965)
26. Morgenstern, L.: Mid-sized axiomatizations of commonsense problems: a case study in egg cracking. Stud. Logica **67**(3), 333–384 (2001)

27. Mösenlechner, L., Beetz, M.: Parameterizing actions to have the appropriate effects. In: IEEE/RSJ International Conference on Intelligent Robots and Systems (IROS), San Francisco (2011)
28. Okada, K., Kino, Y., Inaba, M., Inoue, H.: Visually-based humanoid remote control system under operator's assistance and its application to object manipulation. In: Proceedings of Third IEEE International Conference on Humanoid Robots (2003)
29. Ratey, J.J., Galaburda, A.M.: A User's Guide to the Brain: Perception, Attention, and the Four Theaters of the Brain. Vintage Series. Vintage Books, New York (2002)
30. Reiser, U., Connette, C., Fischer, J., Kubacki, J., Bubeck, A., Weisshardt, F., Jacobs, T., Parlitz, C., Hägele, M., Verl, A.: Care-o-bot® 3-creating a product vision for service robot applications by integrating design and technology. In: IEEE/RSJ International Conference on Intelligent Robots and Systems, 2009 (IROS 2009), pp. 1992–1998. IEEE, Piscataway (2009)
31. Reiter, R.: Knowledge in Action: Logical Foundations for Specifying and Implementing Dynamical Systems. MIT Press, Cambridge (2001)
32. Sakagami, Y., Watanabe, R., Aoyama, C., Matsunaga, S., Higaki, N., Fujimura, K.: The intelligent asimo: system overview and integration. In: IEEE/RSJ International Conference on Intelligent Robots and Systems, vol. 3, pp. 2478–2483 (2002)
33. Saxena, A., Driemeyer, J., Ng, A.Y.: Robotic grasping of novel objects using vision. Int. J. Robot. Res. **27**(2), 157–173 (2008)
34. Shanahan, M.: A logical formalisation of Ernie Davis's egg cracking problem. In: Problem Fourth Symposium on Logical Formalizations of Commonsense Reasoning (1997)
35. Siciliano, B., Khatib, O. (eds.): Springer Handbook of Robotics. Springer, Berlin (2008)
36. Siskind, J.: Reconstructing force-dynamic models from video sequences. Artif. Intell. **151**(1), 91–154 (2003)
37. Tenorth, M., Profanter, S., Balint-Benczedi, F., Beetz, M.: Decomposing CAD models of objects of daily use and reasoning about their functional parts. In: IEEE/RSJ International Conference on Intelligent Robots and Systems (IROS), Tokyo Big Sight, pp. 5943–5949 (2013)
38. Tenorth, M., Winkler, J., Beßler, D., Beetz, M.: Open-ease – a cloud-based knowledge service for autonomous learning. KI – Künstliche Intelligenz (2015)
39. Tulving, E.: Episodic memory: from mind to brain. Annu. Rev. Psychol. **53**(1), 1–25 (2002)
40. Wyrobek, K.A., Berger, E.H., der Loos, H.F.M.V., Salisbury, J.K.: Towards a personal robotics development platform: rationale and design of an intrinsically safe personal robot. In: 2008 IEEE International Conference on Robotics and Automation, pp. 2165–2170 (2008)

Chapter 6
On "Storing Information" in Families: (Mediated) Family Memory at the Intersection of Individual and Collective Remembering

Rieke Böhling and Christine Lohmeier

Abstract When approaching "information storage" from a social science perspective, there are different actors and actor constellations that have a stake in the kind of information that is stored, how it is stored, and therefore in determining what is ultimately remembered. Consequently, a continuous interplay between individual and collective dimensions can be observed in the production of memory in societies. This chapter introduces to (cultural) memory research with a focus on family practices of remembering in today's media environment. We argue that family memory is situated at the intersection between individual and collective memory and can therefore serve to illustrate different strands in interdisciplinary memory research. We first provide a brief overview of memory studies in conjunction with media and then introduce the study of family memory in particular. Secondly, we introduce different ways of researching family memory and illustrate these with a number of examples of empirical studies.

6.1 Introduction

How is information stored in families? At first glance, this appears to be the central question to be discussed in this chapter, which is published in an interdisciplinary volume on information storage; together with chapters ranging from storing information in the brain, to computers, to robots, to artificial intelligence and intelligent systems, and finally to "information storage" in societies from a social sciences and humanities perspective. One can imagine that it is difficult to find common ground

R. Böhling (✉)
Zentrum für Medien-, Kommunikations- und Informationsforschung (ZeMKI), University of Bremen, Bremen, Germany
e-mail: rieke.boehling@uni-bremen.de

C. Lohmeier
FB Kommunikationswissenschaft, University of Salzburg, Salzburg, Austria
e-mail: christine.lohmeier@sbg.ac.at

C. S. Große, R. Drechsler (eds.), *Information Storage*,
https://doi.org/10.1007/978-3-030-19262-4_6

between all of these disciplines but after several meetings with the interdisciplinary working group "information storage" a number of central themes that are of interest in all these disciplines were carved out and we will address several of them later on in this chapter.

However, to begin with, we would like to put forward a "disclaimer," in order to promote a better understanding of the point of view taken in this chapter. When approaching memory in families, we cannot technically speak of "storage" of "information." Instead of referring to "storing" something we much rather can speak of "remembering" something, or "memory"—as these words can imply already that what is remembered is not static or fixed but instead fluid and subject to change, depending on interpretation. In other words, "[r]emembering is not the re-excitation of innumerable fixed, lifeless and fragmentary traces. It is an imaginative reconstruction, or construction, built out of the relation of our attitude towards a whole active mass of organized past reactions or experiences" (Frederic Bartlett, one of the "pioneers of the psychology of remembering" as quoted in [26, pp. 238–39]). Moreover, we encounter some difficulty with referring to "information" when thinking of family memory, as the word "information" can imply some kind of static knowledge—for example, with reference to processes of computer architecture and artificial intelligence, as is discussed in other chapters in this volume (but where complete accuracy is also not necessarily needed and some data may be approximated, cf. Chap. 4). We do not only consider information in a static sense but we consider memories, which emerge through interactions with stored information (of which the level of accuracy can likewise be debated) and memories include feelings, emotions, and stories—which can change over time and are not fixed but fluid and, above all, are difficult to grasp.

Coming from a disciplinary background in social sciences and humanities (and working within the discipline of communication and media studies), we research the social construction of the social world of individuals, in and with media without taking a media-centric perspective [29]. That is, we do recognize the importance of media in everyday communication but we do not place the study of mediated communication above all else. Instead, we are interested in the life worlds of individuals in society, and in this case the ways in which their memories are communicatively constructed within everyday practices.

Thus more precisely—taking into account the disclaimer above as well as our disciplinary backgrounds—the central question of this chapter is: How do families remember in today's media environment? This is of course a very broad question that cannot be answered fully. Instead, we pose the question to unveil certain patterns and describe processes of remembering. This chapter will provide a concise overview of theoretical advancements and methodological approaches in researching practices related to family memory and illustrate this through drawing on case studies and examples. We will pay specific attention to intersections between individual and collective memory.

6.2 Background: From Memory to Family Memory

6.2.1 Collective and Cultural Memory

Memory studies is a growing field that is situated in multiple disciplinary contexts: ranging from the humanities (e.g., literary studies and film studies) to the social sciences (such as sociology and psychology). In 2016, the Memory Studies Association was founded to facilitate exchange between researchers from different disciplines [58]. Despite this richness in different disciplinary approaches, there are nevertheless two shared common threads from the 1920s often cited with regard to collective memory in particular: Maurice Halbwachs and Aby Warburg [13, p. 13, 27, p. 80].

Whereas German art historian Warburg focused on recurring visuals in his ideas on a common visual memory or iconic memory (Bildgedächtnis), French sociologist Halbwachs engaged with memory within collective groupings [13, 27]. Halbwachs [20] postulated that memory is constructed through and within social groups providing frames for the interpretation of past events (see also [34])—a common thread that also features prominently when researching family memory and a point that we will return to below.

Since Halbwachs and Warburg there have been a multitude of studies on memory and different ways of collective remembering (or, more recently, also "collaborative remembering" to emphasize the collaborative character with which remembering takes place; [41]) and similarly there is a large number of different conceptualizations of memory. Beginning in the 1980s, French historian Pierre Nora published several volumes on lieux de memoire—translated into English as realms of memory [comparable to research on memory places (Erinnerungsorte) in German memory research (cf. [17]; cited in [27])], which have been very influential for memory studies. In the introduction to the English edition, Nora states:

> If the expression lieu de memoire must have an official definition, it should be this: a lieu de memoire is any significant entity, whether material or non-material in nature, which by dint of human will or the work of time has become a symbolic element of the memorial heritage of any community [...]. [47, p. xvii]

Nora thus placed an emphasis on cultural memory's connection to communities' heritage and subsequently identity. Focusing on the case of France, Nora introduced a myriad of realms that have become part of France's memorial heritage and form a part of the country's national identity.

Next to Nora, Aleida and Jan Assmann are influential scholars in the field of memory studies in Germany in particular [13, 27]. Among other things, Assmann and Assmann coined the term communicative memory in distinction to cultural memory. Whereas communicative memory is located on the level of the social and "lives in everyday interaction and communication and, for this very reason, has only a limited time depth which normally reaches no farther back than 80 years, the time span of three interacting generations" [2, p. 111], cultural memory is the way in which this memory is institutionalized and made "by means of things meant as reminders such as monuments, museums, libraries, archives, and other mnemonic institutions" [2, p. 111]. Of course, archives, libraries, and the like can be considered as spaces that hold, save, and preserve information.

Moreover, in Germany, the Collaborative Research Center 434 Memory Cultures (Erinnerungskulturen) at the Justus-Liebig-Universität Gießen (funded by the German Research Foundation (DFG) between 1997 and 2008) has also contributed significantly to the development of research on memory cultures (see also [13, pp. 34–37]).

Astrid Erll has emphasized the enormous importance of media for any kind of collective memory, as well as the role of media in the construction of collective memory [13, p. 123] and Martin Zierold underlines that further empirical research is needed on the processes of remembrance under the conditions of modern media systems [64, p. 406]. To be sure, media play a crucial role in everyday life and in "shaping our understanding of the past" [14, p. 3] and many of the more recent publications deal with the different aspects and intersections between media and memory (e.g., [10, 11, 18, 19, 25, 26, 46]), as well as with ways of researching these [28]. Memories are mediated and mediatized[1] in multiple ways and family memory combines both public and private aspects of this process.

6.2.2 Family Memory

As has been postulated by Halbwachs [20], memory is constructed within social groups, which provide the frames of interpretation for past events. The family is one significant social group to provide such frames of interpretation [16, 54, 56] and individuals construct parts of their memory of public (e.g., national or global) and individual (biographical) events with reference to the family. Undoubtedly, definitions of what constitutes a family can differ (on the level of the individual but also from a larger, societal perspective) and are also subject to continuous change (see also [34, pp. 2, 11])—an aspect that needs to be taken into consideration when researching family memory.

The construction of memory takes place within different processes, for example through conversations [62], or in interaction with news reports and journalistic content [12, 63]. These processes include different kinds of artifacts, "memory objects," that are located at the intersection between public and private spaces. A memory object can be any kind of object that might be significant to telling a family's story, visual, literary, or symbolic: such as drawings, photographs, manuscripts, diaries, letters, home videos, broadcasting footage, monuments, or places. The social world's deep mediatization [9] brings about an increase in ways to

[1] On the difference of mediation and mediatization: "While 'mediation' refers to the process of communication in general—that is, how communication has to be understood as involving the ongoing mediation of meaning construction, "mediatization" is a category designed to describe change. [. . .] Mediatization reflects how the overall consequences of multiple processes of mediation have changed with the emergence of different kinds of media" [8, p. 197].

handle such memory aids not only materially but also digitally and it simultaneously enables access to a vast amount of such records. This can change the way individuals communicate about memories and it also shapes the way we remember and forget [61].[2]

In relation to the construction of family memory, "memory work" is particularly helpful as an analytical concept. Memory work refers to the purposeful examination of the past, as Annette Kuhn explains, it is "an active practice of remembering that takes an inquiring attitude towards the past and the activity of its (re)construction through memory" [33, p. 303]. Memory work is a process that is not simply taking place in one's mind but it is an activity that takes place in interaction with (mediated) memory objects, such as films, images, and representations. Memory work "presents new possibilities for enriching our understanding [...] of how we use films and other images and representations to make our selves, how we construct our own histories through memory, even how we position ourselves within wider, more public, histories" [31, p. 46]. Along the same lines, mediated memory work refers to "bundles of bodily and materially grounded practices to accomplish memories in and through media environments" [36, p. 778]—it entails "purposeful, memory-related practices that enact instantiations of personal or collective memories through a wide range of historically divergent and culturally heterogeneous practices" [36, p. 779].

Families conduct memory work in very different ways and (mediated) memory objects as well as mediated representations play a role in this process. As Joanne Garde-Hansen argues,

> [w]e understand the past (our own, our family's, our country's, our world's) through media discourses, forms, technologies and practices. Our understanding of our nation's and community's past is intimately connected to our life histories. Therefore, mediated accounts of wars, assassinations, genocides and terrorist attacks intermingle in our minds with multimedia national/local museum exhibits and heritage sites, community history projects, oral histories, family photo albums, even tribute bands, advertisement jingles and favourite TV shows from childhood. [18, p. 6]

Family memory is therefore always located at the intersection of individual and collective as well as between private and public processes of remembering. And the boundaries between the public and the private are not as clear-cut as it might seem initially. Indeed, even a "private" memory object, such as a family photograph can be contextualized into a broader frame and hold relevance for historical interpretations, for example when it is used in a museum or circulated online ([43, p. 244]; see also [39]). Today's changing media environment contributes to the blurriness of distinctions

> between media involving few in concentrated production catering for many in widespread consumption, the media conventionally associated with the term 'mass', and media involving few in production and few in consumption, the media conventionally associated with vernacular culture. [...] Today this distinction remains important but is increasingly blurred. [48, p. 37]

[2]For a more elaborate version of these explanations on family memory please see [34].

When researching family memory, this distinction (including its blurriness) is especially prevalent, due to different layering of individual and collective memories intertwining.

Lastly, remembering family histories is a popular activity and it becomes most striking and important for individuals when there is a sense of memories becoming lost. Dutch author Elma van Vliet has published a number of books that "do not tell a story but instead ask questions" (from van Vliet's website: "Een boek dat niet vertelde, maar vroeg" [60]) about family members and friends. On her personal website, she explains that the idea for the first book emerged when her mother became seriously ill and the author felt that many questions that she had for her mother were still unanswered so that she decided to make a book for her, entitled *Mam, vertel eens* (English translation: Mom, tell me more). From this, a series of books emerged with invitations to recount memories to mothers, fathers, grandparents, children, siblings, and friends. Moreover, van Vliet developed a number of quiz games that "provide stimuli for conversation and help to retrieve real memory treasures."[3] Some of the questions in this quiz reveal the deeply mediatized character of family memories: What is the favorite movie of the person sitting across from you? Which television series did your mother like to watch in the past? Which movie should everyone have watched, in your opinion? Which event in the news of this year has made a particular impression on you? (*"Welchen Film mag die Person, die dir gegenübersitzt, besonders gerne?," "Welche Fernsehsendungen schaute deine Mutter früher am liebsten?," "Welchen Film sollte deiner Meinung nach jeder einmal gesehen haben?," "Welches Ereignis aus den Nachrichten dieses Jahres hat besonders Eindruck auf dich gemacht?"* [15]).

6.3 Central Questions

Keeping in mind the literature review and current state of research as presented above, the broad questions we research relate to the ways in which families remember—and how "memory work" is conducted. Subsequently, central questions in the field of family memory in communication and media studies and beyond—and specifically in relation to "storing information" are: How is memory selected, structured, organized? How is information made accessible and connected? Who decides and makes a choice of what types of information are stored within the family or another collectivity? And furthermore: What kind of information is stored for the individual memory of a person and what kind of information is stored for the entire family and future generations? What types of events, happenings, and encounters is

[3]From the backcover blurb of the German edition of the quiz: "The Erzähl-mal!-family-quiz provides stimuli for conversation and helps retrieve real memory treasures" ("Das Erzähl-mal!-Familienquiz bietet wertvolle Anregungen zum Gespräch und hilft, wahre Erinnerungsschätze zu heben").

information kept on? What kind of information is overtly and explicitly saved and stored and what is hidden, secret, or implicit? Where is information kept and stored? Where are memory objects kept? How is information passed on and transmitted to the next generation or to other family members? What is lost and what is purposely hidden? In which contexts does the transmission of information take place? How is learning about family traditions facilitated? Which role do special events and ordinary day-to-day life play in this? What is the significance of technological advancements and the change of available technology?

6.4 Contribution: Formal Explanations, Experiments, Examples

This section introduces theoretical and methodological approaches in researching (media) practices related to family memory, illustrating with examples of current and past projects. The field of memory studies is characterized by an immense variety of approaches and perspectives, even within the somewhat narrower boundaries of communication and media research. This holds true for research on family memory as well. In the following, we aim to show the breadth of work focusing on family memory and information storage.

6.4.1 Researching Family Memory in Media and Communication Studies

With qualitative social research in general, the aim is to investigate the lives of people "from the inside" ("*von innen heraus*", [44, p. 10]). Similarly, ethnographic research seeks to find out "what is going on in the social world" and "why people do the things that they do" [38, p. 1]. In media and communication research, qualitative and ethnographic research contributes to knowledge on the variety of roles that media practices play in the lives of individuals. In this respect, we would like to mention again the non-media-centric perspective and Nick Couldry's emphasis on a paradigm that "treats media as the open set of practices relating to, or oriented around, media," where one "decentre[s] media research from the study of media texts or production structures and [...] redirect[s] it onto the study of the open-ended range of practices focused directly or indirectly on media" [6, p. 117]. Similarly, Couldry explains elsewhere that a practice approach considers "how media are put to use in, and help to shape, social life and how the meanings circulated through media have social consequences" [7, p. 8]. In order to gain knowledge on the communicative construction of memory within families, this perspective is certainly useful. When conducting qualitative research, researchers need to actively reflect upon the process of gathering data. As US anthropologist

Clifford Geertz, who is well known for his promotion of "thick descriptions"—that is, describing the life worlds of individuals in as much detail as possible—when researching cultures explained, "[w]hat we call our data are really our won constructions of other people's constructions of what they and their compatriots are up to" (Geertz, 1973, p. 9, quoted in [45, p. 62]). Fien Adriaens [1, pp. 185–86] points out that different traditions exist within qualitative social research. On the one hand, researchers have an understanding of qualitative research as "explicitly political" with the intention "to transform the world with its practices" and on the other hand, researchers take a "more pragmatic approach to qualitative research, which sees it as an extension of the tools and potentials of social research for understanding the world and producing knowledge about it" [1, pp. 185–86].

But how can we reconstruct these constructions of people and what they and their peers are up to, as has been outlined by Geertz? One type of qualitative approach is to conduct qualitative (in-depth) interviews or group discussions in order to gain access to individuals' own descriptions, not necessarily aiming for generalizations but for in-depth descriptions. We refer to the term group discussion in order to emphasize the distinction from focus groups and group interviews. The latter are mainly conducted in a group setting in order to save time when gathering data (cf. [51, p. 304]). Individual interviews and group discussions and the data gathered in these can be amended to designs of more ethnographically inspired research, for example by creating "accumulated ethnographic miniatures" or by performing "virtual ethnographies" ("*akkumulierte ethnographische Miniaturen*," "*virtuelle Ethnographien*" [3, p. 191]). In order to illustrate how studies in the field might look like, in the following subsections, we will introduce a number of studies that made use of these methods in a variety of different ways.

6.4.1.1 Interviews

In the article "Bereavement photographs as family photographs: Findings from a phenomenological study on family experience with neonatal end-of-life photography," published in *Communications*, Sara Martel [40] approaches family memory through exploring neonatal end-of-life photography (photos that are taken in the hospital around the death of newborn children). She conducted interviews with parents who have lost a child of less than four weeks of age and have had these pictures taken, in order to find out how these photographs are used within a family setting and beyond. She reports that the majority of the ten participants in her study did not want to "'keep secret' or 'hide' the fact that they had a child who died" but instead wanted to use the "photographs to make their babies known socially" [40, p. 315]. Martel analyzed her interview material using a phenomenological approach, and in her article, she discusses neonatal end-of-life photography as one kind of domestic photography and its importance for the social reproduction of family life. Lastly, Martel points to tensions that can arise and are emphasized through social media and online sharing opportunities of this type of domestic photography: she finds that there is a culture of invisibility around reproductive loss and mourning that is highlighted by domestic photography's location at the intersection of private and public.

Also relying on interviews as the primary method for gathering data, Donell Holloway and Lelia Green [24], in "Mediated memory making: The virtual family photograph album" display mediated memory-making in everyday life. Using a domestication of technology approach [55] to the social networking site Facebook, they analyze by means of critical discourse analysis a set of data (transcripts and field notes) that they gathered from interviews carried out with children and their parents. They identify Facebook as domestication of technology along the characteristics that were put forward by Silverstone et al. [55]: appropriation, objectification, incorporation, and conversion. Holloway and Green highlight other potential pitfalls of remembering "publicly" on social media (next to the ones mentioned above that were identified by Martel as well as by Thimm and Nehls further below): ownership and privacy.

Another study that made use of qualitative interview data is "Negotiating family history: Media use among descendants of Danish Nazis" by Ole Krogsgaard [30]. Krogsgaard has made an explorative study, during which he conducted five interviews with descendants of former Danish Nazis. His focus is on the ways in which media texts about the Nazi era are incorporated into his participants' construction of family memory. He researches this by means of a "remembered media repertoires approach." In particular, he highlights the negotiation processes at play between the individuals' family memory and the mainstream cultural memory.

However, when working with qualitative interview data and researching memory it is vital to be aware of the "argumentative and context-bound nature of memory" and the fact that "attitudes and recollections of the past generated in interviews can be influenced by the course of the interview conversation" [42, p. 72], as Sabina Mihelj shows in her discussions of the nature of interviews in memory research. In the chapter, which is published in a volume on research methods in memory studies, she demonstrates by way of examples from research conducted in Nova Gorica, a town near the Italo-Yugoslav (today Italo-Slovenian) border, how interviews about the past generate data on two levels:

> On the one hand, every interview account of the past is a product of actual exchanges between the interviewer and the interviewee; on the other hand, it also embodies a set of positions taken in response to the different ways of talking about the past circulating in the broader political, social and historical context. The awareness of this double-layered conversation can be usefully embedded both in the design of the interview questions and in the process of analysis. [42, p. 64]

In her example, she shows this by way of the use of the ideologically contested term "Iron Curtain." Mihelj suggests that such "semantically loaded, controversial terms and expressions for past events and processes" [42, p. 64] can be systematically involved in the interview, in order to perceive of the interview as a "double-layered conversation." This means taking into account both the conversation between interviewer and participant and "the broader 'conversation' with alternative ways of talking about the past circulating in a specific political, social and historical context" [42, p. 74]. Researchers have to be keenly aware of these layers in order to productively conduct fieldwork in contested fields.

6.4.1.2 Group Discussions

As mentioned above, another way to research the construction of memory in the life worlds of individuals is through group discussions. Generally and in opposition to interviews, group discussions are able to capture a collective opinion, reproduced in interaction (cf. [49, p. 88, 51, p. 305]). Particularly for family memory, this is certainly an asset of conducting group discussions because the researcher is able to observe directly how memories are discussed within the collective of the family. In addition, group discussions can also contribute to fruitful knowledge about memory when strangers are asked to discuss their family memories. Annette Kuhn conducted a workshop entitled "Eye to Eye" at a British Council Conference in 2004, with thirteen participants from very diverse backgrounds who were asked to bring their own family photographs for discussion. She explains:

> In most societies, family photographs have considerable cultural significance, both as repositories of memory and as occasions for performances of memory. A study of these processes can be helpful towards understanding how the personal and the social aspects of remembering interact in various cultural settings. [32, p. 284]

In this example, the researcher is thus able to observe how memory work is performed by individuals in a group setting. It is important to keep in mind the nature of memory as non-static but instead fluid and shifting and that therefore group interactions might influence the significance of some aspects of memories over others.

Another example of memory and negotiations and performances of memory in a group setting is illustrated by Rothberg and Yildiz' article on "Memory Citizenship: Migrant Archives of Holocaust Remembrance in Contemporary Germany," in which they introduce a project by the "*Neuköllner Stadtteilmütter*," a group of women with a migration background in Berlin's neighborhood Neukölln. In this project, the women came together in various instances to learn about National Socialism in Germany. This project, which was not specifically designed as a group discussion with the aim to gather data on constructions of memory, illustrates in a more naturally occurring setting how memory is performed in a group. In this case, memory is performed "multidirectionally"—recognizing the "productive interaction between the legacies of different histories" [50, p. 38]. These women with a migration background engaged with Holocaust remembrance, "a history and memory of which they are ostensibly not a part and about which they are frequently said to be indifferent" [50, p. 34]. In Rothberg and Yildiz' reading, this shows memory's multidirectionality—how it can be local and transnational and transcultural at the same time.

6.4.1.3 Other Examples (Interpretative Approaches, Quantitative Approaches)

A different example of researching family memory is Peter Lunt's [37] article "The media construction of family history: An analysis of 'Who do you think you are?'," which is an analysis of an episode of the BBC television program *Who do you think you are*. In this analysis, Lunt read the program's quest for family memory as a combination of different conceptualizations of memory: personal memory, social memory, and cultural memory. Lunt's focus is on the first episode of the program, starring Bill Oddie and he analyzed the form and content of the program in order to draw conclusions on the narrative construction of family memory as performed in the television program by Bill Oddie.

Another approach that aims to show the digital patterns of family memory on the platform Instagram is presented in "Sharing grief and mourning on Instagram: Digital patterns of family memories" by Caja Thimm and Patrick Nehls [59]. They performed a content analysis of 449 user postings (selected by means of specific requirements out of 2790 postings) on the social networking site Instagram, which were marked with the hashtag "*Beerdigung*" (the German word for funeral). The content was coded according to visual messages conveyed in the photos. In their analysis, Thimm and Nehls focus mainly on one category, "selfie of the user" in order to show the "types of visualization of the self in the mourning process" [59, p. 336]. Subsequently, they conceptualize the acts of sharing grief and mourning on Instagram as participation in online mini-publics and they point to a downside of the sharing of grief and mourning online as some of the users have been subject to abusive commentary. This points once again to the tensions that come with sharing the personal online, which has also been observed in Martel's and Holloway and Green's research.

The diversity in approaches briefly presented here serves to illustrate that there are a wide variety of ways to research family memory in communication and media studies. The issues addressed in these studies introduced above relate to specific aspects of family memory and provide explanations toward the communicative construction of memory in their case-specific settings. The following sections will introduce another ongoing project and its design in more detail.

6.4.2 *(Mediated) Memories of Migration*

One of the authors of this chapter, Rieke Böhling, is currently working on a dissertation project that seeks to examine how persons with a "migration background" remember the migration histories of their families and how this memory is constructed. In other words, the aim of the research is to examine how the participants conduct memory work with regard to their families' migration histories. The project takes the so-called European refugee crisis as an empirical access point into the renegotiation and interpretation of (mediated) memories related to migration. The underlying assumption is that memories of migration play a part in contemporary debates on immigration, and are reinterpreted in current social and political debates. The following section will explain in more detail the design of the project and some impressions from the fieldwork, which is still ongoing.

At the core of the research is an ethnographically inspired approach to memories of migration—memories that are, in fact, mediated in multiple ways. Firstly, these memories are mediated in such a way that they are not personally experienced but transferred in the context of the family. This relates to Marianne Hirsch's notion of postmemory, which refers to the second generation of Holocaust survivors [22, 23]. It

> characterizes the experience of those who grow up dominated by narratives that preceded their birth, whose own belated stories are evacuated by the stories of the previous generation shaped by traumatic events that can be neither understood nor recreated. [22, p. 22]

Although in Hirsch's work the term is used only in relation to the children of Holocaust survivors, it can, as she suggests, also be used in other contexts of memories of the second generation in relation to traumatic experiences and events. It has in fact been reappropriated for other contexts already (e.g., in a special issue on postmemory with regard to second generation memories of military dictatorships in Latin America, see [4]) and will be further considered in this project to show its relevance also for the memory of individuals with a migration history in the third generation. Secondly, the described project seeks to examine the public dimensions of mediated representations of migration in the past (in the context of the present) by employing a cross-media approach, not focusing on a single medium only but considering the entire media environment. Employing this cross-media perspective aims to take into account the reality of the current media environment and the way individuals interact with it [35, 52, 53]. Thirdly, the project takes into account the role of personal and private (mediated) memory objects for the construction of memory, also employing a cross-media perspective. In this respect, the concept of memory work is particularly relevant. Memory work in Annette Kuhn's conception—as has been explained above—refers to the purposeful examination of the past and mediated memory work pertains to the accomplishment of memory work "in and through media environments" [36, p. 778]. This investigation places memory work at the core of the (mediated) remembering of migration in order to find out how persons with a "migration background" remember the migration histories of their families and how this memory is constructed.

In particular, the focus is on persons from the "third generation" with a Turkish background, residing in Germany. However, it needs to be noted that in the course of the fieldwork so far, it has turned out that speaking of the third generation can sometimes be difficult as generations are not as clear-cut on the ground as it seems in theory. In this project, third generation means that at least one grandparent of the participant came to Germany as a "guest worker" (*Gastarbeiter*) from Turkey. Germany's guest worker recruitment started in 1955 and Turkish guest workers were recruited from 1961 onward, when the recruitment treaty with Turkey, the so-called *Anwerbevertrag*, came into force [57, p. 11]. The ban on recruitment, *Anwerbestop*, in 1973 halted further recruitment, due to the oil crisis in Western Europe [21, p. 384]. The guest worker background of the participants may take different shapes and forms—participants may have (had) grandparents on the paternal or maternal side who came to Germany as guest workers.

To answer how the family migration histories are remembered, the project takes into account three interrelated dimensions—it examines how (mediated) memory work is accomplished through and with media and media practices in (1) public media representations, (2) personal (mediated) memory objects, and (3) in a situation of postmemory, where the so-called persons with a migration background have not migrated themselves and therefore their memories of the migration history within the family are mediated in multiple ways and on multiple levels. In all of these dimensions, a complex interplay between both collective or public memories, as well as individual or personal memories can be observed.

The project is designed as an ethnographically inspired, multilayered qualitative analysis. This analysis focuses on the personal experiences and interpretations of (mediated) representations of the past in relation to personal memories through fieldwork among third-generation descendants of Turkish "guest workers" in Germany. The first step of this fieldwork is to conduct interviews that include several interrelated dimensions: family histories and memories, personal histories and memories, media and current migratory movements. This part of the analysis gives access to both public mediated representations of the past as well as personal (mediated) memory objects. The participants contribute these public and personal dimensions of remembering through the interviews.

Moreover, this project contributes to further developing and empirically researching the notion of postmemory and tests the applicability of the concept to immigrants' descendants' memories of migration—a migration process that they themselves have not been part of but with which they will nevertheless be confronted in their everyday lives.

Finally, the project draws on and further contributes to researching empirically the existing theoretical considerations on the topics of memory work, as has been explained above. Additionally, José van Dijck describes memory work as a "complex set of recursive activities that shape our inner worlds, reconciling past and present, allowing us to make sense of the world around us, and constructing an idea of continuity between self and others" [11, p. 5]. Considering how memory work is accomplished in families with migration experiences is promising in this respect.

As fieldwork is still ongoing,[4] this contribution can only report on impressions and ideas. In general, in the course of the interviews, it stood out that memory work is accomplished to very different degrees in the families of the participants. Nevertheless, there are some narratives that recur in many of the stories. These narratives are related specifically to the importance of education and obtaining a

[4]At the point of writing, ten qualitative interviews have been conducted. The duration of the individual interviews is between 40 and 120 min.

better life. Moreover, the aspect of forgetting (see [5] for a taxonomy of forgetting and its relevance for memory studies) also plays a part in the narratives: what is deliberately forgotten, what is simply not considered worth remembering? As Paul Connerton notes, much research has focused on memory and remembering and forgetting seems to have a negative connotation, associated with failure, "[y]et forgetting is not always a failure, and it is not always, and not always in the same way, something about which we should feel culpable" [5, p. 59]. This is an aspect worth considering in the field of remembering migration, where different actors have different interests to remember—or to forget—information.

6.5 Future Prospects

Family memory's location at the intersection between individual and collective memory provides a way into conceptualizing the communicative construction of memory. As we have shown in this chapter, a wide variety of approaches toward researching family memory exist. Empirical projects address different aspects of family memory, even within the relatively narrow boundaries of media and communication studies. What has become apparent is that the types of information and the objects that hold information can vary widely. Information for memory work can come from various sources, such as an orally told story, a single photograph, a library, or an archive.

In this chapter, we have first provided an overview from memory to family memory, presenting some ideas from the beginning of memory studies to family memory within media and communication studies. Secondly, we have introduced in more detail ways of researching family memory in media and communication studies, also without taking a media-centric perspective. We have illustrated these ways of researching with a number of empirical studies that made use of different methods. In the end, we introduced one project design, of a project that is still in progress and for which fieldwork is still being conducted, in more detail. This project employs a cross-media perspective, which is promising in particular for the study of memory, as individuals do not make use of only one single medium as a mnemonic device. As we have shown with the example project design, different aspects of the communicative construction of memory can be considered, including digital, analogue, and material aspects of (mediated) memory.

References

1. Adriaens, F: On Mute: Diaspora Youth. A Transdisciplinary and Multi-Method Audience Study into Using, Consuming and Doing Television among Diaspora Youngsters. Universiteit Gent. http://lib.ugent.be/fulltxt/RUG01/001/949/536/RUG01-001949536_2013_0001_AC.pdf (2012)

2. Assmann, J.: Communicative and cultural memory. In: Young, S.B., Nünning, A., Erll, A. (eds.) Cultural Memory Studies: An International and Interdisciplinary Handbook, pp. 109–118. De Gruyter, Berlin (2008)
3. Bachmann, G., Wittel, A.: Medienethnographie. In: Ayaß, R., Bergmann, J. (eds.) Qualitative Methoden der Medienforschung, pp. 183–219. Verlag für Gesprächsforschung, Mannheim (2011)
4. Blejmar, J., Fortuny, N.: Introduction. J. Roman. Stud. **13**(3), 1–5 (2013). https://doi.org/10.3167/jrs.2013.130301
5. Connerton, P.: Seven types of forgetting. Mem. Stud. **1**(1), 59–71 (2008). https://doi.org/10.1177/1750698007083889
6. Couldry, N.: Theorising media as practice. Soc. Semiot. **14**(2), 115–132 (2004). https://doi.org/10.1080/1035033042000238295
7. Couldry, N.: Media, Society, World: Social Theory and Digital Media Practice. Polity Press, Cambridge (2012)
8. Couldry, N., Hepp, A.: Conceptualizing mediatization: contexts, traditions, arguments: editorial. Commun. Theory. **23**(3), 191–202 (2013). https://doi.org/10.1111/comt.12019
9. Couldry, N., Hepp, A.: The Mediated Construction of Reality. Polity Press, Cambridge (2017)
10. van Dijck, J.: Mediated memories: personal cultural memory as object of cultural analysis. Continuum. **18**(2), 261–277 (2004). https://doi.org/10.1080/1030431042000215040
11. van Dijck, J.: Mediated Memories in the Digital Age. Cultural Memory in the Present. Stanford University Press, Stanford (2007)
12. Edy, J.A.: Troubled Pasts: News and the Collective Memory of Social Unrest. Temple University Press, Philadelphia (2006)
13. Erll, A.: Kollektives Gedächtnis und Erinnerungskulturen: eine Einführung. Metzler, Stuttgart (2005)
14. Erll, A., Rigney, A.: Mediation, Remediation, and the Dynamics of Cultural Memory. Walter de Gruyter, Berlin (2009)
15. Erzähl Mal! Das Familienquiz: Verlagsgruppe Droemer Knaur. http://www.droemer-knaur.de/buch/8305112/erzaehl-mal-das-familienquiz-elma-van-vliet (n.d.). Accessed 14 Dec 2017
16. Fivush, R.: Remembering and reminiscing: how individual lives are constructed in family narratives. Mem. Stud. **1**(1), 49–58 (2008). https://doi.org/10.1177/1750698007083888
17. Francois, E., Schulze, H.: Deutsche Erinnerungsorte. 3 Bände. 1. Aufl. C.H. Beck, München (2001)
18. Garde-Hansen, J.: Media and Memory. Media Topics. Edinburgh University, Edinburgh (2011)
19. Hajek, A.: In: Lohmeier, C., Pentzold, C. (eds.) Memory in a Mediated World. Remembrance and Reconstruction. Palgrave Macmillan, London (2015)
20. Halbwachs, M.: On Collective Memory. University of Chicago Press, Chicago (1992)
21. Hartnell, H.E.: Belonging: Citizenship and Migration in the European Union and in Germany. https://doi.org/10.15779/Z38G93H (2006)
22. Hirsch, M.: Family Frames: Photography, Narrative, and Postmemory. Harvard University Press, Cambridge (1997)
23. Hirsch, M.: The Generation of Postmemory: Writing and Visual Culture After the Holocaust. Columbia University Press, New York (2012)
24. Holloway, D., Green, L.: Mediated memory making: the virtual family photograph album. Communications. **42**(3), 351–368 (2017). https://doi.org/10.1515/commun-2017-0033
25. Hoskins, A.: Media, memory, metaphor: remembering and the connective turn. Parallax. **17**(4), 19–31 (2011). https://doi.org/10.1080/13534645.2011.605573
26. Hoskins, A., Tulloch, J.: Risk and Hyperconnectivity: Media and Memories of Neoliberalism. Oxford Studies in Digital Politics. Oxford University Press, Oxford (2016)
27. Jacke, C., Zierold, M.: Gedächtnis Und Erinnerung. In: Hepp, A., Krotz, F., Lingenberg, S., Wimmer, J. (eds.) Handbuch Cultural Studies Und Medienanalyse. Springer, Wiesbaden (2015)
28. Keightley, E., Pickering, M. (eds.): Research Methods for Memory Studies. Research Methods for the Arts and Humanities. Edinburgh University Press, Edinburgh (2013)

29. Krajina, Z., Moores, S., Morley, D.: Non-media-centric media studies: a cross-generational conversation. Eur. J. Cult. Stud. **17**(6), 682–700 (2014). https://doi.org/10.1177/1367549414526733
30. Krogsgaard, O.: Negotiating family history: media use among descendants of Danish Nazis. Communications. **42**(3), 369–379 (2017). https://doi.org/10.1515/commun-2017-0032
31. Kuhn, A.: Family Secrets: Acts of Memory and Imagination. Verso, London (2002)
32. Kuhn, A.: Photography and cultural memory: a methodological exploration. Vis. Stud. **22**(3), 283–292 (2007). https://doi.org/10.1080/14725860701657175
33. Kuhn, A.: Memory texts and memory work: performances of memory in and with visual media. Mem. Stud. **3**(4), 298–313 (2010). https://doi.org/10.1177/1750698010370034
34. Lohmeier, C., Böhling, R.: Communicating family memory: remembering in a changing media environment. Communications. **42**(3), (2017). https://doi.org/10.1515/commun-2017-0031
35. Lohmeier, C., Böhling, R.: Researching communicative figurations: necessities and challenges for empirical research. In: Hepp, A., Breiter, A., Hasebrink, U. (eds.) Communicative Figurations. Transforming Communications in Times of Deep Mediatization, pp. 343–362. Springer International, Cham (2018). https://doi.org/10.1007/978-3-319-65584-0_14
36. Lohmeier, C., Pentzold, C.: Making mediated memory work: Cuban-Americans, Miami media and the doings of diaspora memories. Media Cult. Soc. **36**(6), 776–789 (2014). https://doi.org/10.1177/0163443713518574
37. Lunt, P.: The media construction of family history: an analysis of 'who do you think you are? Communications. **42**(3), 293–307 (2017). https://doi.org/10.1515/commun-2017-0034
38. Machin, D.: Ethnographic Research for Media Studies. Arnold; Oxford University Press, London (2002)
39. Marselis, R., Schütze, L.M.: 'One way to Holland': migrant heritage and social media. In: Drotner, K., Schrøder, K. (eds.) Museum Communication and Social Media: The Connected Museum, Routledge Research in Museum Studies, vol. 6, pp. 75–92. Routledge, New York (2013)
40. Martel, S.: Bereavement photographs as family photographs: findings from a phenomenological study on family experience with neonatal end-of-life photography. Communications. **42**(3), 309–326 (2017). https://doi.org/10.1515/commun-2017-0036
41. Meade, M.L., Harris, C.B., Van Bergen, P., Sutton, J., Barnier, A.J. (eds.): Collaborative Remembering: Theories, Research, and Applications. Oxford University Press, Oxford (2017)
42. Mihelj, S.: Between official and vernacular memory. In: Keightley, E., Pickering, M. (eds.) In Research Methods for Memory Studies, Research Methods for the Arts and Humanities, pp. 60–78. Edinburgh University Press, Edinburgh (2013)
43. Mihelj, S.: Memory, post-socialism and the media: nostalgia and beyond. Eur. J. Cult. Stud. **20**(3), 235–251 (2017). https://doi.org/10.1177/1367549416682260
44. Mikos, L., Wegener, C.: Qualitative Medienforschung: Ein Handbuch. UTB GmbH, Stuttgart (2005)
45. Moores, S.: Interpreting Audiences: The Ethnography of Media Consumption. The Media, Culture & Society Series. Sage, London (1993)
46. Neiger, M., Meyers, O., Zandberg, E.: On Media Memory: Collective Memory in a New Media Age. Houndmills, Basingstoke, Hampshire, UK. Palgrave Macmillan, New York (2011)
47. Nora, P., Kritzman, L.D.: Realms of Memory: The Construction of the French Past. Translated by Arthur Goldhammer. New York: Columbia University Press. http://faculty.smu.edu/bwheeler/Joan_of_Arc/OLR/03_PierreNora_LieuxdeMemoire.pdf (1998)
48. Pickering, M., Keightley, E.: Memory, media and methodological footings. In: Hajek, A., Lohmeier, C., Pentzold, C. (eds.) Memory in a Mediated World. Remembrance and Reconstruction, pp. 36–52. Palgrave Macmillan, London (2015)
49. Przyborski, A., Wohlrab-Sahr, M.: Qualitative Sozialforschung: Ein Arbeitsbuch. Walter de Gruyter, Berlin (2014)
50. Rothberg, M., Yildiz, Y.: Memory citizenship: migrant archives of holocaust remembrance in contemporary Germany. Parallax. **17**(4), 32–48 (2011). https://doi.org/10.1080/13534645.2011.605576

51. Schäffer, B.: Gruppendiskussion. In: Mikos, L., Wegener, C. (eds.) Qualitative Medienforschung Ein Handbuch, pp. 304–314. UTB GmbH, Stuttgart. http://www.utb-studi-e-book.de/9783838583143 (2005)
52. Schrøder, K.C.: Audiences are inherently cross-media: audience studies and the cross-media challenge. Commun. Manag. Q. **18**(6), 5–27 (2011)
53. Schrøder, K.C.: Communicative figurations and cross-media research. In: Hepp, A., Breiter, A., Hasebrink, U. (eds.) Communicative Figurations. Transforming Communications in Times of Deep Mediatization, pp. 407–424. Springer International, Cham (2018). https://doi.org/10.1007/978-3-319-65584-0_14
54. Shore, B.: Making time for family: schemas for long-term family memory. Soc. Indic. Res. **93**(1), 95–103 (2009). https://doi.org/10.1007/s11205-008-9409-2
55. Silverstone, R., Hirsch, E., Morley, D.: Information and communication technologies and the moral economy of the household. In: Silverstone, R., Hirsch, E. (eds.) Consuming Technologies. Media and Information in Domestic Spaces, pp. 9–17. Routledge, London (1992)
56. Smart, C.: Families, secrets and memories. Sociology. **45**(4), 539–553 (2011). https://doi.org/10.1177/0038038511406585
57. Spohn, M.: Alles getürkt: 500 Jahre (Vor)Urteile der Deutschen über die Türken. Bibliotheks- und Informationssystem der Universität Oldenburg, Oldenburg (1993)
58. The Memory Studies Association: Memory Studies Association. https://www.memorystudiesassociation.org/about_the_msa/ (n.d.). Accessed 27 Nov 2017
59. Thimm, C., Nehls, P.: Sharing grief and mourning on instagram: digital patterns of family memories. Communications. **42**(3), 327–349 (2017). https://doi.org/10.1515/commun-2017-0035
60. van Vliet, E.: Mam, Vertel Eens. ElmaVanVliet.Nl (blog). http://www.elmavanvliet.com (2017)
61. Wang, Q., Lee, D., Hou, Y.: Externalising the autobiographical self: sharing personal memories online facilitated memory retention. Memory. **25**(6), 772–776 (2017). https://doi.org/10.1080/09658211.2016.1221115
62. Welzer, H.: Re-narrations: how pasts change in conversational remembering. Mem. Stud. **3**(1), 5–17 (2010). https://doi.org/10.1177/1750698009348279
63. Zelizer, B., Tenenboim-Weinblatt, K.: Journalism and Memory. Palgrave Macmillan, Basingstoke (2014)
64. Zierold, M.: Memory and media cultures. In: Young, S.B., Nünning, A., Erll, A. (eds.) Cultural Memory Studies: An International and Interdisciplinary Handbook, pp. 399–407. De Gruyter, Berlin (2008)

Chapter 7
Cultural Memory and Screen Culture

How Television and Cross-Media Productions Contribute to Cultural Memory

Berber Hagedoorn

Abstract In the modern, overabundant information landscape, information is accessible on and across multiple media platforms and screens, making television and audiovisual memory ever more available. How do the creative practices of media professionals contribute to cultural memory formation today? What is the role of using audiovisual archives to inform and educate viewers about the past? And how can researchers study these dynamic, contemporary representations of past events, and the contribution of audiovisual sources to cultural memory? In this chapter, I consider how new forms of television and cross-media productions, collected in and distributed by audiovisual archives, affect the medium television as a practice of cultural memory in the multi-platform landscape. I zoom in on the role of creative production practices (so-called screen practices) and their social aspects in the construction of memory, in relation to the increasingly dynamic and multi-platform medium that television has become today, and present a dynamic model for studying contemporary television and screen culture as cultural memory.

7.1 Introduction

Current changes in our modern media landscape, such as cross-media storytelling, online archives, digitization, and niche programming (targeting specific audiences or subgroups) have made television and audiovisual memory ever more available to us. This extensive "archive" of television and audiovisual (AV) history for public consumption has been recycled or repurposed by media makers and memory

---

B. Hagedoorn (✉)
Research Centre for Journalism and Media Studies, University of Groningen, Groningen, The Netherlands
e-mail: b.hagedoorn@rug.nl

C. S. Große, R. Drechsler (eds.), *Information Storage*,
https://doi.org/10.1007/978-3-030-19262-4_7

consumers in a number of creative ways.[1] In this chapter, I will zoom in on the role of creative production practices—so-called *screen practices*—and their social aspects in the construction of memory, in relation to the increasingly dynamic and multi-platform medium that television has become today. In this overabundant information landscape, information is accessible on and across multiple media platforms and screens—hence I refer to it as the "multi-platform era" or "multi-platform landscape." Today, in this multi-platform landscape, this variety of interactions is opened up far beyond the television screen, to other platforms, screens, and users. How do the creative practices of television professionals (including the use of audiovisual archives to inform and educate viewers about the past) contribute to cultural memory formation today? And how can researchers study the selection of these dynamic, contemporary representations of past events, and the contribution of these audiovisual sources to cultural memory? In this chapter, I consider how such new forms of television and cross-media productions representing history, collected in and distributed by audiovisual archives, affect the medium television as a practice of cultural memory in the multi-platform landscape.

7.1.1 Outline of the Chapter

To do so, I reconsider television as a practice of cultural memory, taking the medium's hybridity into account. First, I zoom in on the theoretical concept of cultural memory. Second, I consider television's transformation into a dynamic constellation of screen practices, which includes the circulation of produced content across different platforms and screens. Therefore, I offer a critical rethinking of theoretical concepts connected to the medium—specifically liveness as presence and immediacy, fixity, and flow—to address recent developments in television as a memory practice. Finally, by adopting and expanding Aleida Assmann's model of the dynamics of cultural memory between remembering and forgetting, I present a new model to study television, cross-media, and audiovisual archival sources as cultural memory, which takes the medium's hybridity in the multi-platform era into account.

7.2 Cultural Memory

Astrid Erll describes cultural memory not as the object of one single research field or academic discipline, but fundamentally as a "transdisciplinary phenomenon" and

[1]This article is based on a part of chapter 7 from my dissertation: Hagedoorn [25]. The model of television as a hybrid repertoire of memory and connected reflections were previously introduced in: Hagedoorn [23].

"interdisciplinary project." Erll therefore concludes that a favored standpoint or approach for cultural memory research does not exist [16]. Memory studies is a diverse research field where the notion of "cultural memory" distinguishes itself from the concepts of collective memory, popular memory, social memory, and *lieux de mémoire*. Rather, cultural memory is a dynamic practice or constructive process, with a specific focus on the interplay of present and past in sociocultural contexts [17]. Instead of placing the emphasis on sites of memory as relatively stable references for personal and collective memory, cultural memory research today focuses more on how the active relation between present and past is reproduced and how stories are (re-)remembered. Media are assigned a central role in this process, as research by Erll and Ann Rigney among others makes evident [21].

Cultural memory can thus be seen as the complex ways in which a culture remembers.[2] Television programs and related cross-media content, reusing audiovisual and previously broadcast materials, underline how cultural memory is not oppositional to the discourse of official history, but "entangled" with history [50]. As Mieke Bal has stated, the notion of cultural memory has displaced and submerged the discourses of individual (psychological) memory and social memory. This specific term now signifies that memory can be understood as a cultural phenomenon, as well as an individual or social experience:

> The memorial presence of the past takes many forms and serves many purposes, ranging from conscious recall to unreflected re-emergence, from nostalgic longing for what is lost to polemical use of the past to reshape the present. The interaction between present and past that is the stuff of cultural memory is, however, the product of collective agency rather than the result of psychic or historical accident. [...] [C]ultural recall is not merely something of which you happen to be a bearer but something you can actually *perform*, even if, in many instances, such acts are not consciously and willfully contrived [5].

More specifically, cultural memory calls attention to the active, continuous, and unstable process of remembering—and therefore forgetting—in sociocultural contexts [50].

7.2.1 Practices of Memory

The crucial role that media play in the processes of both remembering and forgetting is currently reaching new levels of interest in the interdisciplinary and multidisciplinary study of memory. I advocate a similarly dynamic approach to the study of television today [26]. In the current "multi-platform era," television has become a constellation of dynamic screen practices and can in this manner be studied as a *practice of memory*, following Marita Sturken's understanding of a "practice of memory" as "an activity that engages with, produces, reproduces and invests meaning in memories, whether personal, cultural or collective" [51].

[2] See also: Plate [43].

According to Sturken, the concept of cultural memory is deeply connected to the notion of memory practices, because the active and constructed nature of memory is emphasized. The concept "practice of memory" allows for a focus on television as a *continuous*, *unstable* and *changing* memory practice in the multi-platform era, particularly because the production and reconstruction of memory through cultural practices has as its basis the idea that memories are always part of larger processes of cultural negotiation and transformation. As Sturken argues: "This defines memories as *narratives*, as *fluid* and *mediated cultural and personal traces of the past*" [51]. (my emphasis)

7.3 Rethinking Television Studies

Television's transformation into a constellation of screen practices challenges the dominant conception that television, characterized by liveness, immediacy, and its ephemeral nature, is a disposable practice incapable of memory.[3] Like other media, television has often been theorized as a stable, fixed, and autonomous technology. The medium has also been slated for rendering memory static and enduring. Television's contribution to the loss of historical consciousness has often been attributed to the medium's flow quality. In the present media climate, a critical rethinking of television and theoretical concepts connected to the medium—specifically liveness as presence and immediacy; fixity; and flow—is essential to address the recent developments in television.

7.3.1 Liveness, Presence, and Immediacy

Television has often been regarded as a "bad" memory medium. Television has principally been conceptualized in terms of time, owing to its basic characteristics of liveness and immediacy, but it has been locked in the present tense. According to Mary Ann Doane, the temporal dimension of television is "an insistent 'presentness'—a '*This-is-going-on*' rather than a '*That-has-been*', a celebration of the instantaneous," its own discourse therefore characterized by Doane as "nowness".[4] Being coded as present, immediate, and live, the medium of television has in particular been categorized as amnesic. As Mimi White has argued in her influential essay "The Attractions of Television: Reconsidering Liveness," liveness has principally been used as a key concept for television studies to characterize fundamental

[3]See amongst others: [6, 9, 36].

[4]Doane [10] (For a foundational reading of television's essential liveness, see: Feuer [22]).

ontological and ideological differences between film and television as distinctive media. This has resulted in the outcome that:

> 'Liveness'—as presence, immediacy, actuality—becomes a conceptual filter to such an extent that other discursive registers are ignored. As a result, television's pervasive discourses of history, memory and preservation are too readily dismissed, relegated to secondary status [...] [58].

Through a reevaluation of liveness as television's most definitive ontology and underlying ideology, White has argued that ideas of history and memory are as central to any theoretical understanding of television's discursive operations as ideas of presence, immediacy, and liveness [57]. Critical work that recognizes television's important contributions to memory and historiography is still in the minority, but White's essays have become a prime inspiration for television historians and memory scholars to argue against cultural criticism that characterizes television as amnesic.

For example, historian Steve Anderson has denominated White's work as an important challenge to foundational television theory. In his work, Anderson argues that television has modeled highly creative and stylized modes of interaction with the past, which play a significant role in cultural memory and the popular negotiation of the past [1]. Furthermore, Amy Holdsworth has used White's 2004 essay to argue against the denial of memory as a possibility for the medium.[5] Mari Pajala has also made use of this essay to emphasize how theorizations that position liveness as the privileged form of televisuality fail to explain the persistent interest in memory, history, and preservation on television.[6] These argumentations can be taken a step further by questioning the basic notion of liveness itself as presence and immediacy.

Television criticism has conventionally defined liveness as the medium's main characteristic and aesthetic; however, scholars like Kay Richardson and Ulrike Hanna Meinhof, John Ellis and Paddy Scannell have questioned the "slippery" and "misunderstood" concept of liveness [14, 44, 48]. Television scholars must be careful not to conflate liveness as a technological effect of television—after all, since the 1960s television has predominantly consisted of prerecorded programs. In the words of Ellis:

> The very act of broadcast transmission itself creates a sense of instantaneous contact with the audience. The act of broadcast and the act of witness take place in the same instant, whether or not the events witnessed are taking place 'live' [15].

It is precisely the moment of instantaneous contact that gives television the power to create memory. Work by Anne Wales and Roberta Pearson shows how television

---

[5]Holdsworth's criticism in this context is particularly directed towards Patricia Mellencamp's edited collection *Logics of Television: Essays in Cultural Criticism* (Bloomington: Indiana University Press, 1990); specifically the essays by Mary Ann Doane 'Information, Crisis, Catastrophe', Patricia Mellencamp 'TV Time and Catastrophe, or Beyond the Pleasure Principle of Television', and in a lesser manner Stephen Heath 'Representing Television' and Margaret Morse 'An Ontology of Everyday Distraction: The Freeway, the Mall and Television'. See: Holdsworth [32].

[6]Pajala's criticism in this context is particularly directed towards: [30, 33]. See: Pajala [41].

can for instance function as a facilitator of cultural memory when broadcasting (annual) events of national mourning, commemoration, or celebration [42, 56]. Such broadcasts both actively memorialize, often by using or recycling archival materials for remembrance, *and* create new memories, shaping the viewers' memory of the event as well as television history. In the multi-platform era, the moment of *instantaneous contact* will lie even more in the hands of the television user.

Liveness, presence, or immediacy must therefore not be equated with transiency, and television culture is not necessarily disposable, as Lynn Spigel has also claimed.[7] Television can be considered more in terms of instantaneous contact with the audience rather than liveness, especially since the number of television programs that is experienced out of time by viewers has severely increased in the multi-platform era. This shift in viewing rituals is likely to intensify even more in the years to come. Through instantaneous contact with its audience, practices of doing history on television are an important force in the reconstruction of experiences of the past in the present. What is more, the privileging, marginalizing, and rejecting of certain memory narratives over others by television creators is an important characteristic of the medium as a practice of memory in the multi-platform era.

7.3.2 Fixity Versus Connectivity

In the second place, like other media, television has often been criticized for rendering memory static and enduring. Andrew Hoskins has for example drawn upon research by the neurobiologist Steven Rose to address the acclaimed fixing potential of media, including television and the archive:

> A videotape or audiotape, a written record, do more than just reinforce memory; they freeze it, and in imposing a fixed, linear sequence upon it, they simultaneously preserve it and prevent it from evolving and transforming itself with time [34, 47].

However, Hoskins moves on to argue how "the distinctions between the totalizing and the contextual, the permanent and the ephemeral, the archive and narrative are less effectual when memory is embedded in networks that *blur* these characteristics [and] technological advances that have transformed the *temporality*, *spatiality*, and indeed the *mobility* of memories" [35] (my emphasis). The medium that is of principal interest to Hoskins in this context is the Internet. I propose that televisual practices of re-screening—indicating the vast access to a (digital) repertoire of previously transmitted images in today's multi-mediated landscape [27]—from factual programming to online networked television archives, need to be considered here as well.

[7] Given its ephemeral nature, television is still largely viewed as disposable culture [. . .] [49].

Television has often been theorized as a stable and fixed technology, isolated from other (screen) practices.[8] However, the possibilities of watching television "live" (watching television programs while being broadcast), "near-live" (there is a small time difference between the time of broadcasting and watching a program) and "time-shift viewing" (watching a program recorded at an earlier time)[9] already indicate the versatility of watching television, and exhibit how the dynamics of television as both a practice and experience are constantly shifting. What is more, television programs in the multi-platform era offer additional and connected experiences next to traditional broadcasting, for instance via the Internet, digital thematic channels, and DVD. Derek Kompare has argued that watching a particular text on DVD is a distinct experience from watching that same text on television—or in that respect, in the cinema or on videotape—stating that the DVD box set "functions as a multi-layered textual experience distinct from television and only obtainable via DVD" [38].

I argue that in contrast, such practices must be considered a necessary part of television as a *constellation of dynamic screen practices* in the multi-platform era: in terms of television users interacting with television programs beyond the moment of viewing in different discourses surrounding the television text, but also in terms of collecting and increased personalization, or "Do-It-Yourself" TV archiving. The experience of watching a television series on demand or via DVD in one's own time instead of a weekly broadcast at a set time is also offered via digital thematic channels, on-demand online and streaming services, and time-shifting technologies. This must be considered as one of the many different experiences television currently offers to media users. In this respect, Jane Roscoe has also argued that "choice is the buzzword for broadcasters and audiences" in her discussion of multi-platform event television [46]. Television is constantly connected to other cultural texts and can no longer be considered or theorized as a medium in isolation.

7.3.3 Media Convergence and Flows of Memory

Third and finally, television scholars have generally understood television to obtain its meaning in a manner different to for instance the experience of reading a book, as television presents itself to viewers as a flow of images that can or cannot be related to each other. In the words of Raymond Williams, television's *flow quality* consists of "the replacement of a program series of timed sequential units by a flow series of differently related units in which the timing, though real, is undeclared, and in which the real internal organization is something other than the declared organization" [59]. Work by Williams and Ellis recognized how television

[8]Television is considered by many people to be a stable technology without opportunities for further innovation [40]. See also: The isolated TV set; the picture box cut off from culture [8].

[9]See also: Nikkel [39].

viewers compose their own television text from a variety of segments (in programs, channels, commercials . . .) and how television in this manner can contribute to assumptions, attitudes, and ideas prevalent in a society arising from the ideologies underpinning that society [7, 13, 59]. Television's acclaimed role in the loss of historical consciousness has often been attributed to the medium's flow quality. By implicating flow as an intrinsic quality of television together with liveness, television's contribution to the loss of history was emphasized as a key characteristic of the medium by Stephen Heath, who has argued that:

> The liveness of television—whether real or fictive (liveness is a primary imaginary of television)—also has its significance here, that of a constant immediacy, television today, now, this minute. Exhausting time into moments, its 'now-thisness', television produces forgetfulness, not memory, flow, not history. If there is history, it is congealed, already past and distant and forgotten other than as television archive material, images that can be repeated to be forgotten again [31].

According to William Uricchio, a subtle but important shift in the concept of flow has taken place in the age of convergence, replacing a programming-based notion of flow with a viewer-centered notion of flow and more recently, a new technologically ordered concept of flow [53]. In today's mediated era we can watch television programming via multilayered television sets, personal computers (desktop-, laptop-, tablet PCs) and mobile phones (by receiving either streamed television content via the Internet or terrestrial mobile broadcasting via Digital Video Broadcasting-Handheld (DVB-H)); transmitted via digital and analogue signals; as terrestrial, cable, satellite, handheld/mobile, or Internet television; in or outside the domestic viewing context of the home; in a variety of distribution formats, such as traditional broadcasting, on-demand services, digital thematic channels, DVD productions; and different storage formats, like DVR systems. Uricchio emphasizes that the gradual shift from traditional television broadcasting to alternate carriers and intensified convergence has subsequently granted the Internet access to domains that were once exclusively televisual [54]. This argument can be extended to include other dynamic screen practices as well, especially when considering television's practices of multi-platform storytelling.

Television's convergence with new and digital media technologies has become a distinctive feature of the medium, transforming television from an activity fixed around programming and broadcasting schedules to a practice concentrated around the selection of the television user. As a result, television content, in both Dutch and international contexts, flows across numerous media platforms and screens in a variety of ways. This has also *shaped* television and cross-media creators' *strategies* of repurposing archival materials in new contexts and making history programming accessible across media platforms and screens. This is a specific example of convergence, which Henry Jenkins has defined as "the flow of content across multimedia platforms, the cooperation between multiple media industries, and the migratory behavior of media audiences who will go almost anywhere in search of the kinds of entertainment experiences they want" [37]. The medium television is in itself a unique example of convergence, given that the activity of watching television has become a multi-platform practice. Multi-platform story production and storytelling demands considerable efforts of both creators and users to achieve a deeper engagement. It is a specific mode of engagement and production routine that challenges the use of the medium television. It is also a fruitful line of investigation to gain further insight into television as a constellation of screen practices and a more participatory medium—which involves a set of expectations from creators too. By making television content available on multiple platforms, televisual practices of "re-screening" the past in turn provide television users with an active and continuous link to versions of the past in documentaries and archive-based histories.

The privileging, marginalizing, and rejecting of certain memory narratives over others is an important part of this process. Open to a number of different distribution formats, televisual practices of re-screening consequently produce a *flow*—or indeed, flows—*of memory* through multi-platform storytelling. Instead of each television image replacing the next in a serial succession in television's traditional "flow" model, television images exist continuously side-by-side in a parallel extension on multiple platforms. These images are being navigated through an increasingly viewer-sided and technology-sided notion of flow. Images can be revisited as long as such memory materials keep making themselves available to audiences—which in today's technologically advanced era can both be an exceedingly lengthy period[10] as well as bound by different challenges and restrictions.

7.3.4 New Directions for Studying Screen Culture

Various media ranging from radio, print, online and digital media can work together to provide additional historical frameworks and backgrounds with information provided on broadcast television. It is essential to analyze these strategies as an integral part of television in the multi-platform landscape. By constructing narratives that are too large to be told through one medium, televisual practices of cross-media and transmedia storytelling provide necessary contextual frameworks with televised histories and other representations of the past.[11] Via television as a multi-platform or cross-media experience, viewers can connect with the past on personal, public, national and international levels, demonstrating the continuing importance of stories and memories produced through televisual practices—and challenging accepted versions of history. Television and cross-media professionals working at different levels in the industry have the responsibility to reflect on what kind of representations of the past they give a voice, and scholars should critically assess how this affects the formation of memory in multi-platform environments. For example, without a strategy to *integrally* preserve websites and other cross-media

[10] For a discussion of the possible 'hazards' of the increased digitization of memory, see: Van House [55].

[11] See for example: Hagedoorn [24].

practices with history television programming, important sites of memory will be lost for future remembrance and reflection. Just "because" representations of the past have a social relevance does not necessarily mean that they will be preserved for posterity, or that the forms in which they are offered are suitable to do so [45].

7.4 Television as a Cultural Memory Practice in the Multi-platform Landscape

Television in the multi-platform era, on the one hand, is "adding" more and more cultural artifacts to our cultural history and memory. On the other hand, the reconstruction of memories through practices of doing history is a dynamic process of constant change—rewriting, rejecting, privileging, and marginalizing certain memory narratives over others. Television today functions as a contemporary practice of memory by contextualizing history through a network of dynamic and mediated screen practices—both on the meta-level of television as a multi-platform practice, and on the micro-level of television programs that employ multi-platform storytelling. In this context, I propose a new model (described later on, see Fig. 7.2) to study television and its cross-media content as cultural memory, representing the medium's hybridity in the multi-platform era.

7.4.1 Aleida Assmann's Model of Cultural Memory

Kirsten Drotner has argued that media and memory are "intimately connected" in modern times for the reason that media can not only retain events experiences across time and space, but also help retrieve them at a later date and in another place [11]. Erll in this context makes a heuristic distinction between the three functions media of memory can perform on a collective level: (1) *storage*, as media store contents of cultural memory and make them available across time; (2) *circulation*, since media enable cultural communication across time and space and disseminate contents of cultural memory; (3) as a *trigger* or "*media cue*" for acts of cultural remembrance, and that it is often the narratives surrounding such media or sites of memory that determine their meaning [19]. Aleida Assmann's model of cultural memory (Fig. 7.1) is a crucial instrument here to a deeper understanding of cultural memory as the interplay of present and past in sociocultural contexts and provides further insight into this tension. In this model, Assmann makes an important distinction between remembering and forgetting as both active and passive processes, arguing that "[t]he tension between the pastness of the past and its presence is an important key to understanding the dynamics of cultural memory" [3]. However, the model needs to be reconsidered in the light of contemporary practices of multi-platform television that make evident that cultural memory is increasingly more dynamic. I

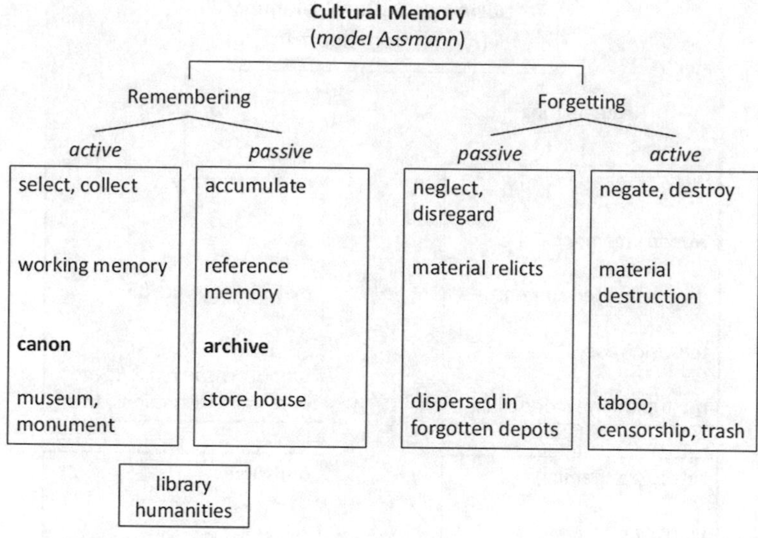

Fig. 7.1 Model Assmann: Cultural memory [2]

take Assmann's model as a starting point for reflection, but I will rework the model based on my own observations of television as a dynamic process and practice of cultural memory in the contemporary media environment.

Assmann has characterized memory as a highly selective practice. Practices of active memory preserve the past as present, whereas practices of passive memory preserve the past as past. Specifically, *actively circulated memory* that keeps the past present is identified as the "*canon*," made perceptible through practices of selection, value, and duration. *Passively stored memory* that preserves the past as past is identified as the "*archive*," denoting storehouses or stable repositories of information and power. The canon can be compared to curated exhibits on display in a museum, and the "archive" to objects hidden from the public's view in the storehouse. The former comprises texts with a sanctified status, destined to be repeated and reread. The latter includes disconnected cultural relics waiting for new interpretations. The cultural practice of forgetting also consists of a more active and a more passive form. A distinction is made between *active* intentional acts of forgetting, like material destruction, and *passive* non-intentional acts of forgetting, such as loss and negligence [18].

7.4.2 Television and Screen Culture Today as Cultural Memory

My model "Television as Cultural Memory" (Fig. 7.2) outlines television as a practice of active and passive remembering and forgetting. In this model, I adopt

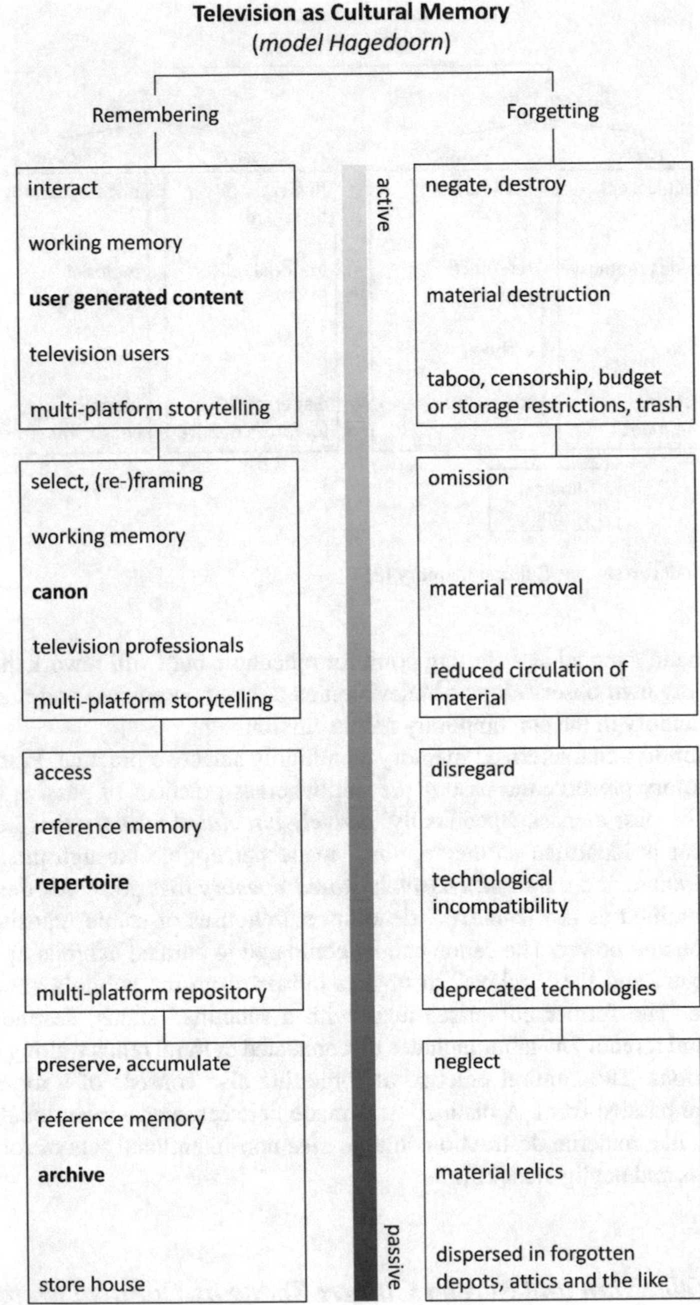

Fig. 7.2 Model Hagedoorn: Television as a practice of cultural memory in the multi-platform era, adopting and expanding Assmann's theory of "canon and archive"

and expand Assmann's theory of "canon and archive" in the context of television. The model makes evident how television as cultural memory offers more dynamic, diverse forms of engagement with the past to different users and in particular, a wider range of opportunities to develop specific memory practices in the multi-platform era.

Assmann's original model needs to be reworked in a number of ways to map out contemporary dynamics. Rather than representing active and passive remembering (or forgetting) on opposite sides of a spectrum, Fig. 7.2 represents *a more dynamic spectrum*. The different levels of active and passive engagement with the past by different users are made visible in vertical relation to one another. I outline different *stages* of remembering and forgetting (from more and most active, to less and least active or passive cultural practices). It is implied in horizontal relations which stage of active/passive remembering is more susceptible to which stage of active/passive forgetting (which does not mean it is invulnerable to other modes of forgetting). In this manner, the model emphasizes the close connections between different forms of remembering and forgetting, and a more nuanced perspective on the degree of disconnection.

7.4.3 The "Working" Memory: Creators' Pre-structuring of Screen Practices

Different user roles in the active construction of a *working memory* are subsequently made explicit, meaning that the role of the television and cross-media professional is emphasized on the level of selection and reframing, resulting in the assembling of content for the canon. The role of media professionals as curators in the construction of narratives of the past in this manner includes the selection and collection of content and researched materials for the canon, but also the reframing and repurposing of broadcast materials on diverse platforms and screens. Television professionals and television users both engage in cultural practices of multi-platform storytelling, contributing to the active reconstruction of memory. I therefore make room for user interaction and the incorporation of user-generated content. Interactive, participatory practices and content produced by television users also need to be considered as a significant part of such a working memory. This is especially relevant considering that television users are more and more becoming like media producers in their own selection of and interaction with content. Such forms of user engagement are *pre-structured* by television platforms as spaces of participation and steered by creators in the way television content is made accessible [29]. Assmann's work has shown that elements of the canon can recede back into the archive, while elements of the archive can be recovered and reclaimed for the canon [4]. In a similar manner, user-generated content can recede into the canon—private memory narratives, audiovisual footage, and comments on television content via social or personal media, to name but a few forms of user interaction.

For example, digital thematic channels show the circulation of televisual content as a practice of cultural memory—for instance, national collective memory as understood by television professionals can inform the scheduling of history programming on the digital thematic channel [28]. This includes a dynamic spectrum of active and passive forms of remembering: from the selection and reframing of memory materials to providing access to a repertoire of connected texts. However, these practices are subordinate to active and passive forms of forgetting. The scheduling and pacing of content for the canon as working memory is mostly subject to more active forms of forgetting, which can include how long a certain program is made available on-demand, how many times a program is allowed to be rerun on a specific digital channel, budget restrictions, copyright issues, and other forms of omission or negation. Forms of user interaction can also recede into the canon, for instance, by television users offering suggestions for documentaries via Facebook. At times content remains accessible as reference memory, receding into the repertoire, for instance when technological incompatibility (as a more passive form of forgetting) impedes the access to a multi-platform repository.

7.4.4 The "Reference" Memory: Archive and Repertoire

In contrast to Assmann's original model, I distinguish between two different forms of *reference memory*. The "archive" functions as the storehouse for accumulation and preservation of audiovisual archival materials and knowledge thereof, including digitization practices and the storing of apparatuses to screen or play particular audiovisual content. However, in the multi-platform era, we can consider another distinct mode of reference memory for television. Diana Taylor has made a useful distinction between the fixed, relatively stable objects in the archive and "the so-called ephemeral repertoire of embodied practice/knowledge (i.e., spoken language, dance, sports, ritual)" [52]. As Erll has also pointed out, Assmann focuses on the uses of mnemonic material, while Taylor draws attention to the specific mediality of such materials [20]. Taylor's definition of the concept "repertoire" alludes more to embodied practices and performances ("... all those acts usually thought of as ephemeral, nonreproducible knowledge" [52]); however, I use the concept repertoire to make visible how television as cultural memory represents a more dynamic form of *access* to televisual content, which is dissimilar from the fixed mode of the archive. Television as a repertoire, then, is not a stable storehouse, but a multi-platform, cross-media repository that is more susceptible and vulnerable to changes over time. This repertoire comprises a wide, variable, and changing range of possibilities to access televisual content across different screens and platforms. The conditions and time constraints under which these materials are accessible to professionals and viewers can vary, and are subject to rights issues and other limits to material circulation. Via new digital technologies, users give active, personal interpretations to multi-platform repositories such as on-demand (online and streaming) services, video-sharing websites, and media platforms.

7.4.5 Active Forgetting

Finally, the model is further expanded by including technological incompatibility as an important form of *disregard*. This is particularly a possibility for the repertoire, which is less fixed and more likely to change or be prone to deprecated technologies in comparison to the archive-as-storehouse. Comparable to Assmann's model, material relics are the most passive form of forgetting as *neglect*, and material destruction is the most active form of forgetting as *negation* or *destruction*. However, material removal is another important form of active forgetting as *omission*. This includes more or less active decisions by television professionals in *not selecting* particular content for the small screen. It also includes the reduced circulation of televisual content on a digital thematic channel due to a limited number of authorized repeats, as well as content made available online for a limited period. Historical narratives and memories transmitted through archive-based and documentary television programs not only represent but also help to preserve the past—which involves dynamic practices of both active and passive remembering *and* forgetting.

Such programming works as a practice of memory and is the end-result of processes of *negotiation* between television professionals. The medium's contemporary dynamics as a textual composite can be further explored. Follow-up research could provide further insight into how television professionals are socialized into the discussed norms and values of doing history; the dynamics of power in the education, learning, and routinizing of necessary skills for doing history on television; how the cultural, textual, and institutional frameworks by means of which the reconstruction of narratives of the past are negotiated and experienced evolve over larger periods of time; further reflections on decision-making processes and organizational constraints for doing history; the power of specific sources; the extent to which professionals in the television industry are guided by similar objectives; and finally, how such practices and processes impact on television as a practice of cultural memory. For such research endeavors, a structural preservation of production research documentation and contextualization materials is necessary—which is as such often not consistently in place—to be able to provide a further understanding of the medium's contemporary dynamics in these contexts.

7.5 Conclusion

Studies of memory comprehend cultural memory as shared and reconstructed knowledge of the past outside of but nevertheless entangled with official historical discourse [43, 50]. New cultures of participation and digital technologies can provide a more direct link between audiences and sources of historical information, but to actively engage television users in spaces of participation, links need to be made meaningful. History television productions and other representations of the past, reusing audiovisual sources, facilitate such negotiations by portraying those parts of the collective memory that are most relevant at the given time to program makers and their audiences.[12] Characterized by a constant process of cultural negotiation, these screen practices and practices of "doing history" reveal the increasingly networked nature of cultural memory. Such practices draw our attention to the mediatedness of memory texts as well as the politics of remembering and forgetting. The reconstruction of narratives of the past through the medium of television is negotiated and experienced within specific cultural, textual, and institutional frameworks, including history, memory, narrativity, medium specificity, house styles, media policy, and contexts of access over time and space. Interpretations are also shaped through viewer expectations and the personal engagement of television users with content, across platforms. Importantly, such experiences are in turn steered by the ways in which content is made accessible by television institutions and media professionals. New digital technologies are the driving force behind these increasingly connected experiences offered and used by the medium television in the multi-platform era.

Television today opens up access to a hybrid repertoire of connected cultural texts made available across multiple platforms and screens. The study of television as a practice of cultural memory therefore not only needs to include the study of memory materials, but also the manner in which this content is curated and made available to the public by television professionals through struggles over power. Reworking Assmann's model of cultural memory based on observations of television as a practice of cultural memory is a step in this direction. The new model emphasizes the interplay of present and past in contemporary televisual environments. Television is being increasingly stylized as a media interface, where the viewer's attention is dispersed across a range of entry points and information triggers. Television as a hybrid repertoire of memory illuminates how texts from the canon may faster recede into the repertoire but also bring about new opportunities to reclaim and contextualize texts for the canon. Fundamentally, television and its cross-media content is a facilitator for the more dynamic ways in which memory content is circulated and made sense of today. Television in the multi-platform era offers a wider range of forms of engagement with the past to different users. These dynamics ultimately make evident the continued relevance of these forms of screen culture and why they should not be forgotten.

References

1. Anderson, S.: History TV and popular memory. In: Edgerton, G.R., Rollins, P.C. (eds.) Television Histories: Shaping Collective Memory in the Media Age, pp. 20–22. University Press of Kentucky, Lexington (2001)

[12]See also: Edgerton [12].

2. Assmann, A.: Canon and archive. In: Erll, A., Nünning, A., Young, S.B. (eds.) Cultural Memory Studies: An International and Interdisciplinary Handbook, p. 99. Walter de Gruyter, Berlin (2008)
3. Assmann, A.: Canon and archive. In: Erll, A., Nünning, A., Young, S.B. (eds.) Cultural Memory Studies: An International and Interdisciplinary Handbook, p. 98. Walter de Gruyter, Berlin (2008)
4. Assmann, A.: Canon and archive. In: Erll, A., Nünning, A., Young, S.B. (eds.) Cultural Memory Studies: An International and Interdisciplinary Handbook, p. 104. Walter de Gruyter, Berlin (2008)
5. Bal, M.: Introduction. In: Bal, M., Crewe, J.V., Spitzer, L. (eds.) Acts of Memory: Cultural Recall in the Present, p. vii. University Press of New England, Hanover (1999)
6. Bertman, S.: Cultural Amnesia: America's Future and the Crisis of Memory, p. 85. Praeger, Westport (2000)
7. Bignell, J.: An Introduction to Television Studies, p. 23. Routledge, London (2004)
8. Caldwell, J.T. (ed.): Televisuality: Style, Crisis, and Authority in American Television, p. 151. Rutgers University Press, New Brunswick (1995)
9. Doane, M.A.: Information, crisis, catastrophe. In: Mellencamp, P. (ed.) Logics of Television: Essays in Cultural Criticism. Theories of Contemporary Culture, p. 227. Indiana University Press, Bloomington (1990)
10. Doane, M.A.: Information, crisis, catastrophe. In: Mellencamp, P. (ed.) Logics of Television: Essays in Cultural Criticism. Theories of Contemporary Culture, p. 222, 227. Indiana University Press, Bloomington (1990)
11. Drotner, K.: Mediated memories: radio, film and the formation of young women's cultural identities. In: Hjarvard, S., Tufte, T. (eds.) Audiovisual Media in Transition, p. 149. Dept. of Film & Media Studies, University of Copenhagen, Copenhagen (1998)
12. Edgerton, G.R., Rollins, P.C. (eds.): Television Histories: Shaping Collective Memory in the Media Age, p. 8. University Press of Kentucky, Lexington (2001)
13. Ellis, J.: Visible Fictions: Cinema, Television, Video. Routledge, London (1992)
14. Ellis, J.: Seeing Things: Television in the Age of Uncertainty. I.B. Tauris, London (2000)
15. Ellis, J.: Seeing Things: Television in the Age of Uncertainty, p. 74. I.B. Tauris, London (2000)
16. Erll, A.: Cultural memory studies: an introduction. In: Erll, A., Nünning, A., Young, S.B. (eds.) Cultural Memory Studies: An International and Interdisciplinary Handbook, pp. 1–15. Walter de Gruyter, Berlin (2008)
17. Erll, A.: Cultural memory studies: an introduction. In: Erll, A., Nünning, A., Young, S.B. (eds.) Cultural Memory Studies: An International and Interdisciplinary Handbook, p. 2. Walter de Gruyter, Berlin (2008)
18. Erll, A.: Cultural memory studies: an introduction. In: Erll, A., Nünning, A., Young, S.B. (eds.) Cultural Memory Studies: An International and Interdisciplinary Handbook, pp. 97–107. Walter de Gruyter, Berlin (2008)
19. Erll, A.: Memory in Culture, pp. 126–129. Palgrave Macmillan, Basingstoke (2009)
20. Erll, A.: Memory in Culture, p. 51. Palgrave Macmillan, Basingstoke (2009)
21. Erll, A., Rigney, A.: Introduction. In: Erll, A., Rigney, A. (eds.) Mediation, Remediation, and the Dynamics of Cultural Memory, pp. 1–13. De Gruyter, Berlin (2009)
22. Feuer, J.: The concept of live television: ontology as ideology. In: Ann Kaplan, E. (ed.) Regarding Television: Critical Approaches – An Anthology, pp. 12–22. American Film Institute, Los Angeles (1983)
23. Hagedoorn, B.: Television as a hybrid repertoire of memory. New dynamic practices of cultural memory in the multi-platform era. VIEW J. Eur. Telev. Hist. Cult. **2**(3), 52–64 (2015). Available online: http://www.viewjournal.eu/index.php/view/article/view/jethc032/60
24. Hagedoorn, B.: Towards a participatory memory: multi-platform storytelling in historical television documentary. Continuum: J. Media Cult. Stud. **29**(4), 579–592 (2015). https://doi.org/10.1080/10304312.2015.1051804. Available online: http://www.tandfonline.com/10.1080/10304312.2015.1051804

25. Hagedoorn, B.: Doing history on television as a practice of cultural memory. In: Doing History, Creating Memory: Representing the Past in Documentary and Archive-Based Television Programmes within a Multi-platform Landscape, vol 15, pp. 162–184. Utrecht, Utrecht University (2016)
26. Hagedoorn, B.: Collective Cultural Memory as a TV Guide: 'Living' History and Nostalgia on the Digital Television Platform. Acta Universitatis Sapientiae, Series Film and Media Studies 14, Histories, Identities, Media, p. 74 (2018). Available online: http://www.acta.sapientia.ro/acta-film/C14/film14-04.pdf
27. Hagedoorn, B.: Collective Cultural Memory as a TV Guide: 'Living' History and Nostalgia on the Digital Television Platform. Acta Universitatis Sapientiae, Series Film and Media Studies 14, Histories, Identities, Media, pp. 76–79 (2018)
28. Hagedoorn, B.: Collective Cultural Memory as a TV Guide: 'Living' History and Nostalgia on the Digital Television Platform. Acta Universitatis Sapientiae, Series Film and Media Studies 14, Histories, Identities, Media, pp. 71–94 (2018)
29. Hagedoorn, B.: Towards a participatory memory: multi-platform storytelling in historical television documentary. Continuum: J. Media Cult. Stud. **29**(4), 590
30. Heath, S.: Representing Television. In: Mellencamp, P. (ed.) Logics of Television: Essays in Cultural Criticism, pp. 267–302. Indiana University Press, Bloomington (1990)
31. Heath, S.: Representing television. In: Mellencamp, P. (ed.) Logics of Television: Essays in Cultural Criticism, pp. 278–279. Indiana University Press, Bloomington (1990)
32. Holdsworth, A.: British Television, Memory and Nostalgia, pp. 31–32. PhD, University of Warwick (2007)
33. Hoskins, A.: New memory: mediating history. Hist. J. Film Radio Telev. **21**(4), 333–346 (2001). https://doi.org/10.1080/01439680120075473
34. Hoskins, A.: Digital network memory. In: Erll, A., Rigney, A. (eds.) Mediation, Remediation, and the Dynamics of Cultural Memory, pp. 92–93. De Gruyter, Berlin (2009)
35. Hoskins, A.: Digital Network Memory. In: Erll, A., Rigney, A. (eds.) Mediation, Remediation, and the Dynamics of Cultural Memory, p. 93. De Gruyter, Berlin (2009)
36. Jameson, F.: Postmodernism, or, the Cultural Logic of Late Capitalism, Post-Contemporary Interventions, pp. 70–71. Duke University Press, Durham (2005)
37. Jenkins, H.: Convergence Culture: Where Old and New Media Collide, p. 2. New York University Press, New York (2006)
38. Kompare, D.: Publishing flow: DVD box sets and the reconception of television. Telev. New Media **7**(4), 335–360, 346, 349 (2006). https://doi.org/10.1177/1527476404270609
39. Nikkel, J., van Bergen, M.: GfK jaargids 2008: uitgesteld televisiekijken. GfK Benelux, Amstelveen and Brussels (2008). http://publications.gfk.nl/?view=jaargids2008.xml
40. Noll, A.M. (ed.): Television Technology: Fundamentals and Future Prospects, p. 1. Norwood, Artech House (1988)
41. Pajala, M.: Television as an archive of memories? Cultural memory and its limits on the Finnish Public Service Broadcaster's Online Archive. Crit. Stud. Telev. **5**(2), 133–145 (2010). https://doi.org/10.7227/CST.5.2.16
42. Pearson, R.E.: Memory. In: Pearson, R.E., Simpson, P. (eds.) Critical Dictionary of Film and Television Theory, p. 388. Routledge, London (2001)
43. Plate, L., Smelik, A. (eds.): Technologies of Memory in the Arts, p. 1. Palgrave Macmillan, Basingstoke (2009)
44. Richardson, K., Meinhof, U.H.: Worlds in Common? Television Discourse in a Changing Europe. Routledge, London (1999)
45. Rigney, A.: Plenitude, scarcity and the circulation of cultural memory. J. Eur. Stud. **35**(1), 11–28 (2005). https://doi.org/10.1177/0047244105051158
46. Roscoe, J.: Multi-platform event television: reconceptualizing our relationship with television. Commun. Rev. **7**(4), 364 (2004). https://doi.org/10.1080/10714420490886961
47. Rose, S.: The Making of Memory: From Molecules to Mind, p. 61. Bantam, London (1993)
48. Scannell, P.: Television and history. In: Wasko, J. (ed.) A Companion to Television. Blackwell Publishing, Malden (2005)

49. Spigel, L.: Our TV heritage: television, the archive, and the reasons for preservation. In: Wasko, J. (ed.) A Companion to Television, p. 92. Blackwell, Oxford (2005). https://doi.org/10.1002/9780470997130.ch5
50. Sturken, M.: Tangled Memories: The Vietnam War, the AIDS Epidemic, and the Politics of Remembering, p. 3. University of California Press, Berkeley (1997)
51. Sturken, M.: Memory, consumerism and media: reflections on the emergence of the field. Mem. Stud. **1**(1), 74 (2008). https://doi.org/10.1177/1750698007083890
52. Taylor, D.: The Archive and the Repertoire: Performing Cultural Memory in the Americas, pp. 19–20. Duke University Press, Durham (2003)
53. Uricchio, W.: Television's next generation: technology/interface, culture/flow. In: Spigel, L., Olsson, J. (eds.) Television After TV: Essays on a Medium in Transition, pp. 171–172, 179. Duke University Press, Durham (2004)
54. Uricchio, W.: Television's next generation: technology/interface, culture/flow. In: Spigel, L., Olsson, J. (eds.) Television After TV: Essays on a Medium in Transition, p. 175. Duke University Press, Durham (2004)
55. Van House, N., Churchill, E.F.: Technologies of memory: key issues and critical perspectives. Mem. Stud. **1**(3), 302–306 (2008). https://doi.org/10.1177/1750698008093795
56. Wales, A.: Television as history: history as television. In: Davin, S., Jackson, R. (eds.) Television and Criticism, p. 53. Intellect, Bristol (2008)
57. White, M.: Television liveness: history, banality, attractions. Spectator. **20**(1), 40–41 (1999)
58. White, M.: The attractions of television: reconsidering liveness. In: Couldry, N., McCarthy, A. (eds.) MediaSpace: Place, Scale, and Culture in a Media Age, pp. 79–80. Routledge, London (2004)
59. Williams, R.: Television: Technology and Cultural Form, p. 93. Schocken Books, New York (1975)

Chapter 8
The Complicated Preservation of the Television Heritage in a Digital Era

Leif Kramp

Abstract Television as an electronic audiovisual mass medium has always been difficult to collect and to preserve. The daily routines of broadcasting institutions have hardly promoted attempts of strategic-objectified collecting. Under the influence of a cultural inferiority of television in general public perception, facilities dedicated to the collection and archiving of this rather new medium have struggled to find ways of coping with its specific challenges. Television as ethereal-volatile as well as material-complex medium seemed to contradict the conventions of institutional archiving that evolved over centuries with a dominant focus on written documents. As an electronic medium, television always depended on technological auxiliaries to endure over time, either by means of film, magnetic tapes, digital versatile or hard discs and playback devices. In the complexity of its material legacy, broadcasting institutions could not always adequately guarantee the collection of televised programming over the past decades. This chapter discusses key issues of television heritage management with a focus on the technical dependencies of long-term preservation of audiovisual broadcasting history. Beyond that, the chapter asks how digitalization promises solutions for the many challenges television preservation faces in the light of material decay and the apparent contradiction between the ostensible omnipresence of media and the ephemerality of broadcasting.

8.1 Introduction

When US President Lyndon B. Johnson delivered a speech to the top representatives of the US broadcasting industry on April 1, 1968, he did not joke when he attested to the audiovisual mass media that they possessed the power of enlightenment, but also that of confusion, since accuracy was all too often sacrificed to the directness of the image. His following sentence, which was broadcast live, did

L. Kramp (✉)
Zentrum für Medien-, Kommunikations- und Informationsforschung (ZeMKI), University of Bremen, Bremen, Germany
e-mail: kramp@uni-bremen.de

C. S. Große, R. Drechsler (eds.), *Information Storage*,
https://doi.org/10.1007/978-3-030-19262-4_8

not get lost in the airwaves: "Unlike the printed media, television writes on the wind" (quoted in [38: 85]). For television archivists, this statement, which is by no means critical of preservation, appears retrospectively to be a reminding appeal to conserve the medium and its contents that Johnson criticizes. The immaterial qualities of television became the biggest challenge for securing cultural assets—and they still are today. In order to continue to serve as an instrument of societal self-understanding, expression, identity formation, and identity maintenance and thus as a tool for social memory work, the televisual audio-visions must be captured and preserved from a normative perspective: a task that television broadcasters were only able to recognize in its full significance and fulfill rather late in the history of television. "Every asset has a life after it goes off television," says Doug Gibbons of the Paley Center for Media,[1] expressing a fundamental wisdom of TV heritage management: Television is not a presentistic act, but remains functional even after it has been broadcast. However, the prerequisites for this were and are only manageable under considerably difficult conditions.

[1]The qualitative in-depth interviews quoted in this chapter were conducted as part of a larger study on television heritage management in North America and Germany, published in the German language in [26]. The sample includes interviews with the following experts and their institutions who are quoted in this chapter (in alphabetical order): Ruta Abolins, Walter J. Brown Media Archives and Peabody Awards Collection (Athens, Georgia, USA); Don Adams, CBC (Toronto, Canada); Daniel Berger, Museum of Broadcast Communications (Chicago, Illinois, USA); Howard Besser, New York University (New York, NY, USA); Susanna Betzel, RTL Television (Cologne, Germany); Thomas Beutelschmidt, Humboldt Universität zu Berlin (Berlin, Germany); Steve Bryant, BFI National Archive (Berkhamsted, UK); Axel Bundenthal, ZDF Television (Mainz, Germany); Kathy Christensen, CNN (Atlanta, Georgia, USA); Glenn Clatworthy, PBS (Arlington, Virginia, USA); Dan Einstein, UCLA Film and Television Archive (Los Angeles, California, USA); Wolfgang Ernst, Humboldt Universität zu Berlin (Berlin, Germany); Laurie Friedman, FOX News (New York, NY, USA); Doug Gibbons, Paley Center for Media (New York, NY, USA); Dina Gunderson, CNN (Atlanta, Georgia, USA); Michael Harms, SWR (Baden-Baden, Germany); Bettina Hasselbring, Bayerischer Rundfunk (Munich, Germany); Hans Hauptstock, WDR (Cologne, Germany); Michele Hilmes, Wisconsin Center for Film and Theater Research (Madison, Wisconsin, USA); Geoffrey Hopkinson, CBC (Toronto, Canada); Chuck Howell, Library of American Broadcasting (College Park, Maryland, USA); Henry Jenkins, Massachusetts Institute of Technology (Cambridge, Massachusetts, USA); Joel Kanoff, ABC (New York, NY, USA); John Koshel, NBC (New York, NY, USA); Heiko Kröger, NDR (Hamburg, Germany); Peter Paul Kubitz, Deutsche Kinemathek (Berlin, Germany); Sam Kula, Consultant (Ottawa, Canada); Bary Lord, Lord Cultural Resources (Toronto, Canada); Mike Mashon, Library of Congress (Culpeper, Virginia, USA); Olaf Moschner, RTL (Cologne, Germany); Horace Newcomb, University of Georgia (Athens, Georgia, USA); Dietmar Preißler, Haus der Geschichte der Bundesrepublik Deutschland (Bonn, Germany); Mark Quigley, UCLA Film and Television Archive (Los Angeles, California, USA); Peter Schwirkmann, Deutsche Kinemathek (Berlin, Germany); Ron Simon, Paley Center for Media (New York, NY, USA); Lynn Spigel, Northwestern University (Chicago, Illinois, USA); Hans-Gerhard Stülb, DRA (Wiesbaden, Germany); Mardiros Tavit, ProSiebenSat.1 (Berlin/Munich, Germany); Lynne Teather, University of Toronto (Toronto, Canada); Robert Thompson, Syracuse University (Syracuse, NY, USA); Gerlinde Waz, Deutsche Kinemathek (Berlin, Germany).

8.2 Background: The Irreversibility of Archival Losses

Those who keep more can remember less: The more extensive an inventory of records is, the more demanding are the challenges of archiving, the higher is the indexing effort, the more likely is the lack of overview, the more difficult is the management and maintenance of the material, the more difficult is the utilization of the fundus. One could conclude: The more archival material, the more important it is to discard documents that are deemed not worthy of archiving in relation to others. On the one hand, however, the so-called cassation of archival documents requires effective, farsighted, and well-founded criteria that limit the possible damage caused by the irreversible destruction of records. Equivalent replacements are usually not available, at least when it is a question of assets that are managed in an archive with a status of an "Endarchiv" (final archive), which is independent of state archives and holds documents that cannot be found elsewhere. The particular responsibility of the broadcasters' archives on this issue was recognized rather late: It is estimated, for example, that only a fraction of US television broadcasting has been handed to archival departments. Horace Newcomb (University of Georgia) assumes a rate of delivery of only 20% from the period from the 1940s to the year 1960. Michele Hilmes (Wisconsin Center for Film and Theater Research) even believes that at most 5% of television broadcasting from the 1950s was preserved, and perhaps 10–12% from the 1960s and 1970s. Mike Mashon from the Library of Congress, however, considers an exact quantification of the actual loss to be pointless, since this does not differentiate between material worth preserving and largely irrelevant broadcasting records:

> I don't know if there are really hard figures on how much TV was not preserved. It's kind of a misleading figure anyway because it would be unfair to say, for example from an 18 hours broadcast day only 30 minutes were preserved what would lead to a really low percentage. One shouldn't think in those figures. (Mike Mashon, Library of Congress)

Nevertheless, the estimates, which undoubtedly vary in detail, point to the same problematic core: the extent of the loss of material has a direct impact on a television company's ability to remember television and its own history. James H. Billington, chief librarian of the Library of Congress, criticized in a preface to a fundamental study on the status of television archiving in 1997 that it is primarily chance that determines which television programs survive. In his view, what remains is an incomplete testimony of the achievements and failures of an entire media culture—with all the adverse consequences for the self-perception of the public community (cf. [7]). Countless landmarks of contemporary history, which has been shaped by television since the middle of the twentieth century, are blurred because at best written documents can still tell about it, but the televisual testimony is lost. The television history of the United States is rich of such gaps: Neither can the first televised address of a US president to the general population in 1947 be found in the archives, as Harry S. Truman appealed to his compatriots to reduce their food consumption for the benefit of hungry Europeans, nor the first prime time news broadcasts of major networks such as the *CBS Evening News*, *All-Star News* on

ABC, or *Camel News Caravan* on NBC until 1955. In addition, there are also gaps in some cases lasting several months with regard to popular shows from the first "Golden Age of Television" such as the *Texaco Star Theatre*. Neither Johnny Carson's debut in his legendary entertainment show *The Tonight Show* nor numerous historically significant productions by the important screenwriter Rod Serling from the early 1950s have been handed over to the archives. These are just a few prominent examples, however, which do not even give any indication of the actual extent of the lost programming assets from US television history:

> Any television historian gets crazy thinking about all that stuff that was lost [...]. Any TV historian can tell you a number of stories like the TV network didn't have enough storage room so they dumped all of those episodes. And that includes not only the tapes and everything, it also includes paper and ephemera. (Robert Thompson, Syracuse University)

In Germany, as archivist Susanne Betzel at the large commercial TV broadcaster RTL admits, the situation of audiovisual preservation in some broadcasting archives is "blatant." Especially from the early years of the Cologne-based private broadcasting group, only limited material was preserved:

> The background is simply that RTL had its beginnings in Luxembourg in the years from 1984 to 1988, so there was no television archive. Of course, many of the materials were then also cherished and cared for by the individual editor and regarded as personal property in part. With the move, criteria were laid down that did not have to be complied with. No one has assessed this, but only whether the material should be taken to Cologne and whether it should be broadcast there. Of course, these were only very short-term decisions. (Susanne Betzel, RTL)

Whereas inconsistent collection activities and errors in the archive evaluation of programming material at RTL have led to several Bundesliga (German soccer league) seasons not being preserved and thus irretrievably lost, there are also some sensitive omissions on the part of public broadcasters: Popular entertainment shows like "Der goldene Schuss" or "Vergiss mein nicht" of the second largest public broadcasting service in Germany ZDF have been lost due to wrong decisions. Other broadcasting archives also complain that in the past it was apparently decided with a false impetus that "such series as 'Dancing lessons with the married couple Fern', 'Gymnastics with Adalbert Dickhut' or 'Gymnastics with Mrs. Pilss-Samek' are not worthy to keep, so that at most one or two archival copies from this sequence of 30 would suffice" (Hans Hauptstock, WDR).

To discuss the large blank spaces in television archiving in order to demonstrate the insanity of many a cassation decision must not, however, hide the fact that the demands on television heritage management have only developed vehemently since the mid-1980s. Archivists always decide in the context of their institutional obligations and can only to a limited extent free themselves from the constraints of their time and the dilemma of assessing historical significance (cf. [26: 122–319]).

8.3 Key Issues

At first glance, it may seem hardly plausible that the creations of such an omnipresent medium as television have disappeared in such an unmanageable and in detail hardly comprehensible number, literally never to be seen again. Ultimately, this also resulted in initiatives to establish television museums in Germany and North America: These were based on the lack of the industry's understanding that too many parts of television history had been lost. Nonetheless, the losses of archival sources can be explained from the context of their time and circumstances. This was usually based on a combination of human error, pragmatic auspices and technological determinants: Specific mindsets, scarcity of resources, loss by usage, and unauthorized action.

8.3.1 *Specific Mindsets*

The so-called deletion trains, which the initiative to establish an independent broadcasting archive and museum entitled "Deutsche Mediathek" (cf. [25]) used as a critical metaphor for the active destruction of programming material in broadcasting archives, were not a regular phenomenon, as respondents in the archive survey emphasize retrospectively. Nevertheless, however, the cassation practices tore deep gaps into the archival fundus, which is increasingly used by a large number of different academic research disciplines as a source for the analysis of contemporary media history. As Heiko Kröger, head of the television archive at the public broadcaster NDR in Hamburg, confirms, it is certainly not wrong to attest that early archival practices at broadcasting institutions were characterized by some form of a throwaway mentality:

> The production idea had priority [...] and there was a lack of understanding for one's own historicality. That, at a later point in time, the repetition of the whole 'Tagesschau' newscast (the 'Tagesschau' 20 years ago is a classic in this respect) would become relevant, no head of programming or production could imagine in the 50s and 60s. (Heiko Kröger, NDR)

A total of three concerted deletion campaigns were carried out by the NDR in the 1970s and early 1980s, by which some broadcasts were almost completely destroyed, as in the case of the show "Music from Studio B": "On both the archival and the programming side, the importance of individual programming elements in terms of their programmatic reinstatement and also in terms of their historical value was misjudged," says Kröger. The fact that this process was not an individual case is obvious because of the reason for the cassations at that time: The surveyed circle of experts repeatedly refers to space problems, costs, and issues of usage with regard to the form of storage and the archive focus, which was oriented almost exclusively toward the short-term reusability of a broadcast or a sequence of it. Also in the United States, this general approach hit the broadcasts from the morning, midday, and afternoon programs with particular intensity because they generally had lower ratings and seemingly a limited lasting value attributed by the program planners, as Ron Simon from the Paley Center for Media explains: “It’s hard to document the everyday game shows or even the everyday newscast of television. That is an example of what we are always looking out for, what daily television looked like.” Laurie Friedman from the Fox News archive also refers to the prevalent production priorities for the evening and prime time programs:

> [W]e didn’t keep a lot of the daytime programming. We kept the primetime shows, some four o’clock forward. But daytime between 9 am and 4 pm we recycled in the early days of the network. That was a collector’s decision: We need tapes back to reuse. The tapes weren’t really being used for the amount of space they were taking up. It’s not at all a big loss for us, because we don’t need it for production. (Laurie Friedman, Fox News)

In principle, the same was true for connecting programming elements in the everyday flow of programming that became more and more important in the course of increasing competition among broadcasters in order to bind viewers to their programming: Trailers and previews, as well as programming notes using announcers’ appearances, were archived just as rarely as sporadically recurring gap fillers, as in the case of the RTL’s mascot “Karlchen” or more regular broadcast pieces such as the weather forecast as curator Gerlinde Waz from Berlin-based Deutsche Kinemathek emphasizes:

> Especially the forms of presentation, as with RTL’s Karlchen, there is at most one example in the archives. Announcers, however, as far as public television stations are concerned, and all those things that are not distinctive broadcasts but rather short linking forms of presentation such as weather, announcements and other elements: If you find something like this in the archives, it is a pure coincidence. (Gerlinde Waz, Deutsche Kinemathek)

The influence of short-sighted production perspectives on the preservation of the television heritage became all the more evident when, in the event of a strained business situation, some smaller broadcasting stations got rid of their complete archive holdings that had been maintained for years, says Sam Kula, archive consultant and former head of the audiovisual archives of the National Archives of Canada:

> I know of many, many cases in North America where television stations had valuable libraries of footage which they built up over ten, fifteen years which was practically a history of the region in which they operated: all aspects of live in the community. But the question was whether the storage room with all the old tapes is really needed. So, in so many cases, archives were wiped out (Sam Kula, archive consultant).

Following the assertions by the surveyed experts, the biggest losses have been experienced in local television. The traditional situation at small television stations is deemed desolate, as scholar Henry Jenkins (Massachusetts Institute of Technology) and archivist Glenn Clatworthy (PBS) point out: Whereas national broadcasting is collected quite comprehensively nowadays because of sufficient financial, personal and material resources, local stations do not have resources alike at their disposal. The audiovisual documentation of television history on a local level especially in the form of local newscasts, is a considerable problem until today:

> These people work under such deadlines, that they are not worried about what they did yesterday and whether it will be saved. It's very hard to even get their ear. [...] It may have an archivist, but many will just call it tape-librarian. They only keep track of their various clip-reels so they can draw them in if they need one. In the future there is no comprehensive effort to save copies of individual news broadcast from every night of the week. They probably have it or at least some of it, but what its long-term future is, is unclear. We try to get to these people, but it is a little bit discouraging sometimes. (Chuck Howell, Library of American Broadcasting)

Even larger central archives and broadcasting collections can rarely replace the losses in such cases, since local television programming is only available in the respective region and does not exceed its narrow geographical boundaries: "Occasionally we will get a call from someone interested in a current local production and they contacted the station who wouldn't have it and they would call us as a place of last resort and unfortunately we don't have any of the local productions except to the extent they entered national distribution. Normally we can't help in that situation," says Glenn Clatworthy from the PBS central archive. Frequently, it is only possible to reconstruct what own productions have been shown on local television after extensive research in press archives. Thus, it is also quite possible that an archival asset actually still exists, but cannot be found, because indexing and cataloguing methods were not used or were used too inconsistently. Since there are no uniform protocols for archiving in television archives in North America or Germany, it takes a lot of patience to achieve results in the search for lost broadcasts: "It can be a real hunt to find something," says Lynn Spigel, professor at Northwestern University. This is also confirmed by John Koshel, archivist at NBC, who uses the example of an iconic program narrative about the resignation of US President Richard Nixon to explain, as the corresponding excerpt shows, how Nixon left the White House when entering the Marine One helicopter with the Victory symbol, was in no way lost, but was reused in a later television report about the inauguration celebrations of the successor Gerald Ford, without it having been noted in the catalogue data:

> So often you would find an archival asset in a later reproduction, e.g. in a highlight-show of the week's news or in an annual retrospective of the news highlights of the year. So often they removed a special piece of film and integrated it into another context. So sometimes the main problem was poor record keeping: When you look for a specific news event and go to the original place in the archive where the reference data guide you, it wouldn't be there. So, people would say: 'What a devastating loss! Why was that thrown out?' But, actually, it only has been moved and moved and moved—because it was reused over and over again over time. (John Koshel, NBC).

8.3.2 *Scarcity of Resources*

Much programming content from the early days of television has been lost, however, not because it was disposed of at a later date, but because, due to the high cost of storage materials, they were not considered for preservation: "So much of the early television was live. There was not much recording made of it. It was screened once and if you were lucky you saw it, and if not, it has gone into space somewhere," says Dan Einstein from the University of California Film and Television Archive. If one refrained from making elaborate film recordings of a live broadcast, it was lost in the airwaves. Sufficient motivation for such efforts was rarely given since the cost–benefit ratio was deemed insufficient due to the poor quality of the programs filmed from the television screen and the reuse value was correspondingly low.

Not affected by this problematic consideration procedure, which turned out to be negative in the question of recording relevance for the majority of programs in cases of doubt, were programming contents that had already been produced in advance and thus recorded on film. This was a costly and material-intensive process, but often not to be avoided within the production process and also, in view of a future reusability, considered to be justified not only for film productions and serial formats, but also for informational programs, especially for news features. This led to a highly fragmented archival fundus that only contained certain parts of a news program that had already been pre-produced as individual contributions on film and those recorded during the broadcasting. Thus, the NDR's programming archive contains a large number of pre-filmed news features from the "Tagesschau" until 1973, but not the word segments in which the anchors read the news, which were no less relevant. Although the preservation of individual news features made it easier to reuse them in other contexts, the complete context of a newscast cannot be reconstructed because numerous, orally presented daily information were not documented in the archive.

With the development of analogue video recording in 1956, archive practice hardly improved at all, but rather the possibility of multiple recording of the magnetic image carrier led to a rapid deterioration of the preservation situation: The higher image quality of the video recordings led to an enormous increase in the attractiveness of recordings for reuse purposes, but the relevant quality of the new technology for the broadcasting industry was the potential for multiple reusability. The result was a rigorous dubbing of recorded programming content as soon as a new broadcast was due to be recorded. In particular, the US-American television industry was blessed with the new technical possibilities, according to Joel Kanoff (ABC): "Programmes were rerecorded because they could be recorded. You couldn't rerecord film." Today, for example, there are often only written traces in the form of clues on the videotape which, however, lead to nowhere:

> Sometimes I get a box with a tape with a game show or something on it. And the log tells you about the forty times the tape was used before. And you read for example: 'Space Shuttle landing' and you can only say: 'Oh, no!' There is an awful lot that's gone. (Dan Einstein, UCLA Film and Television Archive)

Video tapes, as well as their recording and playback equipment, were initially only affordable for large TV companies. An early two-inch quadruplex video tape cost several hundred US dollars at that time—in Germany the prices went into the thousands. In view of the high cost pressure, the regular retransmission of the tapes only corresponded to the production logic, but the archival potential of analogue storage material was hardly recognized until the mid-1980s. As independent archive consultant Jeff Martin stated on the occasion of the 50th anniversary of video technology, the medium did not allow the development of an archive mentality based on the new storage technologies, because it occupied the newly developed memory with a transmission logic: "[V]ideotape was not for the long-term, but only for the now" [31: 58]. According to his study, video tape technology was not invented to serve as a conservation tool, but to reduce the cost of transmission. From the outset, the US television broadcasters had to struggle with four (later five) time zones in order to reach their national audiences. Broadcasts that shimmered live on TV screens on the East Coast were first recorded on film and later on videotapes so that they could be broadcast only a few hours later on the West Coast. The singularity of live television meant that the use and durability of videotapes for distributing the program as many times as possible was in the foreground, and not the longevity and stability in the archive sense. One of the most prominent victims of this overplaying practice was Johnny Carson, who in his popular "Tonight Show" commented on the events of the day in a pointed and ironic way. Whole vintages of his show were not preserved:

> The Johnny Carson show in the 1960s was done on a videotape, and the 2-inch videotape was very expensive and NBC used to re-record over those tapes. So, you have many important historical figures that appeared on that programme, but they didn't realize at that time that anybody would like to watch those old shows again. (Mark Quigley, UCLA Film and Television Archive)

8.3.3 Loss by Usage

Due to the seamless integration of broadcast archives into the production process, the storage locations were always in danger of damaging their storage materials by the active use of the stored programming material for current broadcasting operations or editorial internal screening purposes. As Mardiros Tavit, archivist at German commercial broadcasting group ProSiebenSat. 1, points out, all tape formats are endangered by too frequent playback. The pragmatic approaches that have been pursued in accessing archived video tapes for the longest time and which are still commonplace today in some cases, combined with economic constraints, have prevented a preservation strategy that has the long-term preservation of archival assets as its primary goal. For example, it was and still is part of the standard procedure to give one-off tapes, for which there are no backup copies, out of the hands to serve broadcasting operations. The CBS News Archive has 3000 tapes circulating every week, which are viewed by employees of the various information programs of the broadcaster and copied in sequences as required. In this case, older image carriers in particular run the risk of not being handled properly by inexperienced personnel and thus being damaged (see also [17]):

> We have problems more often with damaged tapes from the ¾-inch era or early betacam-tapes that are deteriorating and we try to train our producers and tell them: 'If the tape doesn't play, then stop and send it back to us.' (Joel Kanoff, ABC)

In view of the high demand from production services, Michael Harms from German regional public broadcaster SWR emphasizes that "by no means we make copies of everything that we exchange in house": "So, these are unique pieces that are exchanged. We only make copies when it's beyond SWR." John Koshel (NBC) underlines this with a vivid example of reporting on the terrorist attacks in New York on September 11, 2001: At that time, the newsrooms only had a single volume of valuable recordings of the attacks, which had to be passed around because there was not enough time to make one, let alone several copies. Already this kind of technically unaccompanied use of an archival document can lead to damage or complete loss, as Mardiros Tavit confirms:

> It has already happened that tapes have been removed and thus irretrievably lost. That could happen again, of course. If the tape cannot be obtained again, there is a certain amount of money that the perpetrator of the damage has to pay as a penalty. (Mardiros Tavit, ProSiebenSat. 1)

Here, the television heritage is characterized as an object of an asset that can be depreciated and compensated for in monetary terms, but whose historical relevance is pushed into the background by administrative imperatives.

8.3.4 Unauthorized Action

According to some of the experts interviewed, however, there have also been conflicting developments, with broadcasters and production companies recognizing the value of their own programs and protecting them from the tireless machine of rebroadcasting. This led to theft of data carriers in numerically undeterminable cases that were removed from the archive records and stored in an office or the private home of the respective person without this intervention becoming known: "People took a lot of stuff home in the old days. And now they have a garage full of old television shows." (Joel Kanoff, ABC)

The situation is similar in sector for written archival assets, where documents from the creative and administrative fields are withheld partly out of negligence, but partly also with a certain intention of archiving. For example, the Historical Archive of the regional public Bavarian Broadcasting Corporation (Bayerischer Rundfunk) was confronted with the problem of the targeted destruction of extensive files prior to the possibility of archival evaluation:

> Before there were any regulations such as the service directive, there were huge losses. Former administrative director Helmut Oeller has retired and simply had his entire stock of

documents destroyed for whatever reason. That's gone, of course. So, sometimes there are whole departments where nothing has been handed down to the archive. We also have very little television from the 1950s. There was simply no awareness of it. (Bettina Hasselbring, Bayerischer Rundfunk).

The loss of internal documents on business conduct and administrative procedures in a television company due to the initiative of individual leaders illustrates the ambivalent relationship between senior television directors who pursue a history policy by disregarding the need for preservation and who are attempting to influence the historical image of a television broadcaster by the extinction of traces of their actions according to their will. The questionability of this approach at public television broadcasters is in no way inferior to the explosive nature of a state archiving activity directed toward political interests, since in both cases it is a matter of documenting the broadcasting process that has shaped society's self-image and would have been able to trace it retrospectively.

In the recent past, there have been some efforts, which were more or less successful, to actively seek untraceable evidence of television programming history. The surprising discovery of extensive film recordings from the first two decades of television operations in Germany among archival holdings of the Deutscher Fernsehfunk (DFF), the television broadcasting organization in former East Germany, which had been made by the GDR leadership for reasons of strategic political observation of the German TV program, had a singular character. There were also numerous complete recordings of "Tagesschau"-editions and broadcasts of the political magazine "Panorama" (NDR). Although these are not available in broadcastable quality, they represent an important scientific resource that helps to document an important part of the broadcasting activities of West German television. The same also applies to film recordings of GDR television by the Federal Ministry of the Interior of the Federal Republic of Germany as well as ARD and ZDF, which had an equal interest in documenting the program events in the neighboring country during the same period. This also meant that the gaps in the older archive holdings of the DFF could by no means be closed; however, the contextual documentation value of the records is immense, as researcher Thomas Beutelschmidt from Humboldt University Berlin puts it:

It is hard to believe what could be reconstructed from this footage, what was recorded, and made available to the German Broadcasting Archive. Above all, this is necessary where they have these mixed programmes, i.e.: You have a short report in a news magazine, then the report is somewhere in the German Broadcasting Archive. But the important thing is that the entire structure of the programme, the moderation of the programme, which was broadcast live, has not been preserved, of course, because the entire programme was not recorded. But due to the Western recording, the overall picture is still there. (Thomas Beutelschmidt, Humboldt University Berlin)

In the international arena, there is considerable hope for private individuals and former employees of the television industry, who may have old tape stocks with their own recordings or professionally produced recordings. While broadcast archives rarely take part in the search for such lost material, the public and scientific institutions in the survey sample understand it as their duty to close collection gaps with their means. The more recent the missing documents are, the greater the likelihood that there will be private recordings of the original program: "There are more and more ways to record programming, not merely by the producing or transmitting organization itself, but there is more and more work that is collected by collectors, by amateurs", says Horace Newcomb, professor at the University of Georgia. The book "Missing Believed Wiped," published by the British Film Institute [18], is an international role model for the attempt to produce a broad public for the problematic situation of the television heritage: On the basis of 21 examples of lost programs from British television history, the book draws attention to the fact that, due to the in part catastrophic situation in independent archives, the examination of television program history can only be practiced with a negligible stock that corresponds to a "tip of the iceberg": "Every genre, indeed every sub-genre, has important gaps among the archive holdings" [18: 139]. Although there are so far few indicators that such a strategy of engaging a considerable, large audience in the search for traces of television programming heritage could be crowned with success, Mike Mashon of the Library of Congress also plays with the idea of initiating a similar television archaeological "national campaign" in the United States: If only a few collectors would report their treasures on it, this could be a gain.

The former Museum of Television and Radio in New York launched such an initiative in cooperation with the cable station Nick at Night (cf. [26: 501–513]) in the 1990s entitled "Lost Programs Campaign" (cf. [41]), which is still being continued today by the Paley Center for Media. The aim of the campaign was and is to create a general awareness of the fatality of the loss of media heritage by means of iconic television broadcasts from national and international television history and to be able to close some of the gaps with the help of private collectors. In this way, it was possible to acquire a complete recording of the 1954 television play "12 Angry Men" (CBS), which was thought to have been lost for decades. Of the one-hour drama, which was broadcast live in the anthology show "Studio One" at that time, only a film recording of the first 30 min was available. Almost 50 years later, the complete recording was discovered by the children of the defense lawyer and later judge Samuel Leibowitz in the private archive of their father, who had received the film reels from the broadcaster shortly after the broadcast because he was interested in the legal questions dealt with in the broadcast (cf. [29]). Another example is that of TV puppet maker Morey Bunin, who died in 1997, whose children donated the original dolls to the American Museum of the Moving Image shortly after his death, but initially considered his 200 film reels of earlier children's broadcasts with popular puppet characters such as "Foodini," "Pinhead," or "The Schnozz," which also appeared, worthless and deposited them in the cellar. It was only through press releases that reported on the great loss of heritage in television history and the resulting high collector's value for original recordings that appeared to be lost that his family became aware of the historical relevance of these traditions and donated parts of the previously carelessly treated collection to the Library of Congress, the Paley Center for Media, and the UCLA Film and Television Archive (cf. [33]).

The role that pure luck can sometimes play in the discovery of old recordings is also evidenced by the reappearance of the first episode of the legendary sitcom

"I Love Lucy" which was celebrated as "Hubble telescope to another time" and aroused great public response [39]: Although the pilot broadcast had never been known to anyone since it had never been broadcast, but was played live as a test balloon and filmed on TV screens inside the studio, its historical value and the great interest shown by a broad public were undisputed in view of the epochal success of the series. Although the film part was initially thought to be privately owned by Desi Arnaz, co-producer and husband of the leading actress Lucille Ball, it was discovered by chance by the widow of the actor Pepito Pérez, who appeared as a clown in that episode, more than 14 years after his death. Pérez had received the film copy from his fishing friend Arnaz as a gift.

Thus, the loss of a tradition in broadcast archives does not necessarily lead to total oblivion in the sense of the uselessness of the lost programming content for cultural memory work, since copies can still be found in attics, cellars, or offices, albeit often in inferior quality. One of the latest discoveries concerns the first "Super Bowl," the infamous championship final of the American professional football league on January 15, 1967, which was broadcast by two of the big networks, CBS and NBC, but none of them considered it necessary to record and archive it. After almost 40 years, a collector offered the Paley Center an almost complete recording of the show [16].

In Germany, on the other hand, there are no comparable initiatives by television broadcasters or public institutions to fill gaps in the inventories of traditional material with the help of the general public, which leads to this, that the problem of the loss of recordings and written documents was only addressed temporarily by isolated press publications such as in the course of the debate on the development of the "Deutsche Mediathek" initiative, but that it is unlikely to be present in the public consciousness, which is also reflected, among other things, in the low interest of the broadcasting archives in possible offers by private collectors.

8.4 Present Attitude of Preservation Between Indoctrination and Change of Consciousness

Even in times of a multi-million-dollar retail trade for the commercial exploitation of historical and current TV productions on the video and online streaming market, only little has changed in the fundamental weaknesses of television heritage management: Awareness of the necessity and profitability of the program's archival material also led to an increase in helplessness toward the overwhelming mass of ongoing archival appraisal. In order to master them, some broadcasting archives are forced, despite significant progress in technological development, not to keep archive material permanently available. Even though the elimination of program substances always remains a risk area, as Axel Bundenthal (ZDF) warns, increasing costs for the maintenance of growing archival stocks put the archive departments under considerable pressure to justify themselves vis-à-vis the station management: "[T]here is pressure to not keep stuff," admits Dina Gunderson of CNN, for example. Under the still adverse conditions for the preservation of the television heritage, the pressure to destroy recordings that are only ascribed a short- to medium-term value for the program operations or to neglect them in conservation efforts is still a burden (cf. [26: 202–221]).

In particular, nonprofit institutions with a broad collection focus have a particularly difficult task in securing what is regarded in the spheres of "final archival" competence as bearable, since even professional observers with knowledge of entrepreneurial archival logic find it difficult to overlook where something might be lost: "The situation today is even worse with the expanding number of networks. It becomes impossible for a public institution to save everything that passes over television when you have hundreds of networks broadcasting 24 hours a day," says researcher Henry Jenkins (MIT). Jenkins therefore calls for an independent supervisory authority to intervene before the archive holdings are collected by the broadcasters and production companies and, after a separate evaluation, takes over if necessary: "They hold on to the content as valuable intellectual property in the short term and then they dump it as worthless in the long-term without any ability of public institutions to step up and collect it." However, the broadcasters reject such a (obligatory) public institution: Olaf Moschner (RTL) refers to the entrepreneurial right of self-determination of the broadcaster to dispose of archive holdings in any given case without consulting an external institution. Since broadcasters play their cards close to their chest how they practice their archival competences (cf. [26: 60–69]), the weal and woe of archival appraisal continue to depend on productivity and profitability value of the archived recordings.

Far from saving the television heritage in a reasonable breadth, the appreciation of the commercial marketing potential of certain parts of TV production has in no way led to a generalized preservationist zeal in the television industry. Rather, the judgments of the archive administrations have been sharpened for a lucrative focus on restoration, that is, the rescue of individual collections. The successful sitcom "Seinfeld" illustrates how the new preservation principle works: For the technical preparation of the historical program material for high-definition television, an unprecedented effort was made. In a three-and-a-half year restoration process, the 180 episodes of the series were completely reworked for several million US dollars on behalf of the production company Sony Pictures Television in order to adapt them to the current technical quality requirements and market them profitably (cf. [24]). However, some of the experts interviewed see—all positive effects of such an enterprise for the preservation of relevant television productions aside—a risk of dominant market orientation in favor of short-term profit goals with only a negligible fraction of the existing archival assets to be considered for such costly restoration projects: "[T]hose entities have other means of distribution that they see as a revenue stream and they keep it in house rather than giving it to someone where it is safe. So, the long-term preservation is endangered" (Mark Quigley, UCLA Film and Television Archive). Also Robert Thompson (Syracuse University) and Daniel Berger (Museum of Broadcasting Communication) do not believe that a capitalist archival ideology could save the television heritage: The focus of the efforts to market archival material as quickly and profitably as possible is too much on the idea of profit. Accordingly, the probable consequences are customized expenses for deferred heritage areas which long-term safeguarding can be regarded as assured in any case due to its popularity and its repetitive qualities in current program planning. Rather, according to the critics, this has fostered a development that increasingly no longer grasps television productions in their programming context, but "recovers the narrative and aesthetic context of a programme once it has been broadcast, which indexes thematically relatively independent visual material" [35: 86], describing it as an autonomous entity without any reference to its originating background and broadcasting environment:

> In the 1950s or 60s television programmes were syndicated via kinescopes or 16-mm prints that had the national network commercials embedded in prints. A lot of the archival material we have from that period has those national spots in it. So, if you want to see a programme in context with its sponsorship and how that might have impact on the programme, it's an opportunity to see some of that, while now if current television is being exploited on DVD, we are not acquiring material recorded off air, we are not likely to get material within the context of its broadcast (Mark Quigley, UCLA Film and Television Archive).

This applies to fictional productions such as TV films and series as well as sports events or shows, which, without the advertising blocks and fade-ins that are often so organically implemented during broadcasting, convey a completely different picture than in their original broadcasting context. In order to obtain a relatively complete impression of the past program events, an imaginary feat on the part of the viewer is often required, since the individual program sections are archived separately and often incomplete at different locations: "When you write about television, you have to piece together a mosaique of what might have actually been on air," reminds researcher Lynn Spigel (Northwestern University).

The fatal dependency of the archive efforts of private production companies on business success has an aggravating effect, which means that long-term preservation cannot be guaranteed due to entrepreneurial imponderabilities alone, and individual archival stocks are thus exposed to unforeseeable dangers:

> It is certainly often difficult for private production companies, and this was also an issue in the debate on the EU Convention on the Protection of the Audiovisual Heritage, where there is even more work being done on producers, so that certain things are not kept permanently in the respective station. Of course, this creates a problem if the production company becomes insolvent or if it is taken over by a new production company. I don't want to deny private companies the consciousness or effort, but the other form of production makes preservation even more difficult. (Axel Bundenthal, ZDF)

Far more devastating is the situation of television heritage management at smaller regional and local stations. It is widely unclear who takes care of the rich as well as confusing archival fundus at those companies that is latently at risk of being lost. Small broadcasters often maintain only a rudimentary archive structure which, for reasons of cost, personnel and space as well as legal restrictions, are often unable to fulfill their functions as a local documentation agency of public life in the service of the community's general interest (cf. [40]). "We still run into situations where they want to get rid of their archived materials because they need or don't have the space", says archivist Sam Kula. Such television broadcasters, but also production companies, continue to find it largely impossible to manage the proper archiving of their programs, which can sometimes be expressed in a weak preservationist attitude:

> [L]ocal television stations are notorious for having their material thrown out. [. . .] Many small TV-stations can't afford an archive [. . .] Take a local station like WALB: We got their newsfilm collection some years ago and the station manager there who was very concerned of the material, signed off the rights and simply said: Take the stuff. Then a different station manager comes in who knows that there is another section of these films, and he doesn't care where they go. They could go anywhere. They just didn't care. (Ruta Abolins, Walter J. Brown Media Archives and Peabody Awards Collection)

On the one hand, there is a gap between an increased professional archive awareness of financially strong broadcasters who have learned from their past mistakes and for reasons of a "burned child effect" (Axel Bundenthal, ZDF), as well as indoctrinable pragmatists, mainly in the archives of commercial broadcasters, who follow a primarily economic impetus and fail to recognize the cultural-historical significance of archival loss. On the other hand, local and regional television broadcasters have only limited financial resources, and it can be seen that a weakly developed network with other broadcasters and archive specialists and the consequent lack of specialist knowledge often stifle the emergence of conservation. This deprives an important but, due to its close local focus and the already devastating state of local television archiving, a part of the television heritage that has generally fallen out of the focus of historiography of adequate care and measures for long-term preservation, which, as will be shown in the following, are associated with immense challenges.

8.5 Measures for Long-Term Archiving

Preserving cultural heritage as long as possible is the central problem of all memory organizations that are committed to preserving and making use of testimonies from the past. When it comes to the long-term preservation of digitized archival assets, an unspecific time horizon is always used, but this should be—from a normative perspective—extended to a maximum. The Canadian museologist Cameron Duncan, in a fundamental article on the ideal conditions for preserving the cultural heritage, set out a number of requirements that an exemplary conservation institution must meet: The air inside the building must be free of pollutants, furthermore it be completely dark, the room should have a constant temperature not lower than 15 and not higher than 20 °C, the relative humidity must not exceed 50–60%, and the structure of the building must provide protection against pressure and sound waves. In addition, Duncan recommends that the cultural asset is sealed off from all organisms—and here mankind must also be expressly involved—as well as a building location in the highlands and with fireproof sheathing. Sophisticated emergency and control systems are just as self-evident as a good connection to the Almighty [9: 17]. As theoretical as Cameron's sanguine checklist may be, it also implicitly draws attention to the fact that the administration of the cultural heritage can only ever fulfill its primary duty to secure historically significant and memory-relevant traditions with inadequate means.

Nevertheless, some countries do not shy away from the effort involved in ensuring the best possible conditions for the long-term preservation of their cultural assets in order to protect the most precious works of their history. In 1974, for example, Germany set up a "Central Rescue Site" in the Black Forest, where the "most important documents of German history" (see [37]; cf. also [32]) are kept in the so-called Barbara tunnel at the foot of the Schauinsland mountain under the auspices of the Federal Ministry for Civil Protection and Disaster Relief and, significantly, not by the Federal Commissioner for Culture and the Media. Securing cultural assets with a time horizon of more than 500 years is limited here exclusively to photographed documents transferred to microfilm. Audiovisual recordings are completely excluded.

8.5.1 In the Race Against Time: Materialized Television Between Decay and Format Clutter

While traditional storage media such as paper or microfilm, despite certain restrictions, may easily prove to be resistant to ageing over hundreds of years, audiovisual media content challenges the existing preservation strategies, as their storage media prove to be far more fragile. Long-term preservation, this is pointed out by the surveyed experts, is simply considered impossible: Pampered by the storage primus paper, conservable due to its wooden basis for centuries, the audiovisual media archives have to find out these days that the program memory they manage, that is, the entirety of their archived heritage, is built on an already heavily undermined cliff that is threatening to break up. Below—to stick to the not groundlessly dramatic image—lies only the ocean of oblivion, because, after all, deterioration of the storage medium could lead to considerable losses. However, as early as the end of the 1980s, the instability of the tape-based data carriers became evident (cf. also [19]). The pressure to develop one's own strategies in order to be able to cope with the threatening challenges weighs heavily on the still relatively young cultural institutions and, last but not least, on the broadcasting archives, all of which have fallen behind in the development of possible solutions due to the already diagnosed lack of understanding of the historicity of the supposedly so presentistic medium of television.

There are no reliable figures on the total number of endangered stocks in television archives and could not be more than indicative figures, since it is not possible to provide reliable information on the extent to which ageing damage threatens the archival inventories in question. The largest library in the United States is also struggling against the clock: only 5% of the audio and video material threatened with decay, according to the administration of the Library of Congress, could have been preserved with the available funds until 2015 if it had not been possible to establish the National Audiovisual Conservation Center (NAVCC) by means of a public–private partnership agreement (cf. [26: 112–117]) and thus to preserve at least half of all endangered stocks for long-term preservation. The NAVCC is one of the few facilities in the world where film and video tape stocks can be preserved at precisely matched temperatures and humidity, thus protecting against premature aging. The fact that the construction of the new archive center has an exceptional signal character, especially with regard to the European archive infrastructure, is underscored by the rarity value of an eight-digit investment sum solely for the conservation measures of the audiovisual heritage, which otherwise only represented an additional and, until not long ago, a particularly annoying side business: "We were able to design this building around our workflows," says Mike Mashon of the Library of Congress.

Despite these groundbreaking improvements in the working conditions of television program preservation, archivists are particularly concerned about the need to cope with storage technologies that do not allow for reliable long-term preservation or the development of satisfactory concepts to control their fragility. The overwhelming majority of program recordings are stored on video tapes of different formats. Even experienced television archivists and video engineers cannot say how long they will be readable: if some old collections are still usable after 50 or 60 years, in some places there have been difficulties in playing back tapes after just 10 years. Among the institutions surveyed, the rough rule of thumb is that video tapes of any format enter a critical phase after about 20 years in which it is no longer possible to ascertain whether they can still be played back without damaging the content stored on them: "You can't guarantee anything, but we store our material in a temperature and humidity-controlled storage facility. Today [we] tried to transfer a ¾-inch tape and it didn't work, even although it was stored properly and was not older than 20 years," says Dan Einstein of the UCLA Film and Television Archive.

> There is the prospect of certain tragedy as we 'go to the videotape' only to find our history reduced to snow on a flickering TV screen. Whether the loss of memories exists on the scale of a single life or that of an entire nation, the loss of primary documents recorded on videotape threatens our cultural legacy. [8: 3]

The most serious problems are currently encountered with magnetic tape formats, which over the decades have proven to be a cost-effective and space-saving alternative to archiving film material. Although film as a storage medium is still considered to be the most reliable carrier, it is also considerably more cost-intensive due to the high material and space requirements compared to video tape formats, while at the same time having to cope with higher demands on the climatic regulation of the archive environment. While the majority of fictional and commercially exploitable television productions are recorded on film and archived accordingly in the production and broadcasting companies, the overwhelming remainder of the television heritage is available in a large number of different archived video tape formats. Preventive conservation measures in the form of optimized storage conditions were taken—if at all—only at a late stage, so that film material in television archives today can become slightly brittle and cracked in the case of dry gluing or elastic in the case of wet gluing and can thus easily become damaged and unusable. Last rescue attempts by so-called recanning, in which the old film rolls are taken out of their metal cans, chemically treated, and then preserved in new metal containers, require specially equipped rooms with special equipment for treating the unstable materials, which, however, hardly any of the transmitter archives can show. With video tapes, in turn, there are repeated shifts of layers that can ruin the video tape and the playback device in equal measure. More problems are demagnetization, oxidation, contamination, wear out, and tape salad.

However, the archiving conditions in most institutions are still not ideal today. While new facilities such as the NAVCC and modern extensions such as at the UCLA Film and Television Archive are meticulously designed to ensure that climatic conditions are maintained, most broadcast archives have to make do with the in-house facilities provided by corporate management. In addition, there are also considerable differences between industrial and nonprofit organizations in dealing with the obvious symptoms of the crisis in the fight against the material decay of television broadcasting between them: While some broadcasting archives do everything in their power to save the most profitable and most requested assets in their possession, and spare no expense or effort, some of the museum facilities rely almost entirely on external preservation efforts by broadcasters for lack of resources.

Peter Schwirkmann from the Deutsche Kinemathek is optimistic: “If something breaks down, we can retrieve it from the archive at any time.” Curator Don Adams of the Canadian CBC Broadcasting Museum takes a similar view: “[Long-term stability] doesn’t matter for us, because in the case that we can’t use one of the VHS copies any more we go back upstairs and get a new master-copy.” Confidence in the preservation expertise and efforts of the archival departments of the television companies leads to a lasting relationship of dependency, which Doug Gibbons of the Paley Center for Media wants to avoid for his institution and pleads for independent efforts for long-term preservation. Apart from the NAVCC, the UCLA Film and Television Archive, and the Paley Center for Media, however, hardly any other institution can afford to take care of the long-term preservation of its collections on master tapes to a similar extent. This raises the question of who is responsible for the long-term archiving of television heritage when the multitude of small and medium-sized cultural and research institutions cannot meet the demanding conditions of preserving audiovisual memories. As researcher Michele Hilmes puts it:

> Here we are: The work of art in the world of mechanical reproduction: Walter Benjamin. What is the TV artifact? First of all there is the whole problem with formats. I am constantly reminded by our archivists that DVD is not a stable format. Videotape is not a stable format. Film is a more stable format. Does that mean that every museum should own 35-mm films of TV-shows? It’s produced on that. But I don’t think so. There should be only a few archives and museums that preserve things. Most of us will be using copies. That is just perfectly keeping with television as a medium. (Michele Hilmes, Wisconsin Center for Film and Theatre Research)

The copyness of television is Michele Hilmes’ argument for the substitutability of its content: A few cultural institutions are concerned about the availability of the unique specimens and take responsibility for the needs-oriented replenishment of copies. In fact, this scenario corresponds to a pragmatic attitude, which assumes that not all memory organizations need to take care of the preservation of the basic fundament of the television heritage, but that they should do better to make a contribution to the exploitation and dissemination of the content: "The largest museum of course can afford to have a large conservation lab, but for the vast majority, even for the large ones, the tendency is to maintain only a minimum of conservation capability in house and to contract out mostly" (Barry Lord, Lord Cultural Resources). From this perspective, it is advisable to focus on the presentation skills of museum institutions in particular, rather than using their limited resources for archival operations. Ruta Abolins (Walter J. Brown Media Archives and Peabody Awards Collection) and archive consultant Sam Kula also argue in favor of developing a partnership between specialized archival institutions and museums in order to preserve material where sufficient resources are available and where it is best known:

> The preservation function is an expensive ongoing operation. It doesn't have to think about all the issues in terms of new formats and changes and constant needs to transfer the collection from one format to another. It could let someone other think about that and focus on the programming and footage it needs. It makes the life of a museum a lot easier to be able to focus on the museology aspects of its work rather than on the archival aspects which are very expensive (Sam Kula, Consultant).

Some non-museal institutions such as the PBS broadcasting archive or the Vanderbilt Television News Archive do not themselves operate long-term backups, but cooperate with the Library of Congress, which functions as their "deep storage": "They send us their digital files, and we tuck them away in our archive. They certainly have them there and they access them, but we essentially do the archiving. We have a lot more capacity that other people ever will to store this material," says Mike Mashon (Library of Congress). In addition to increased capacity, organizations specialized in archiving and restoration of audiovisual storage media generally also have appropriately trained personnel who are entrusted solely with the maintenance and long-term preservation of archival material and are not burdened with other tasks. For the Paley Center for Media, which manages its own collection, the outsourcing of maintenance work is also part of the usual procedure. New York-based VidiPax, the world's largest commercial service provider in the field of magnetic tape restoration, offers support for long-term archiving of audiovisual media in order to recover and back up the stored data. In addition to the Paley Center, customers include all major US networks (except CBS), the UCLA Film and Television Archive, the Library of Congress and numerous other museums, cable stations, film studios, and universities.

The need for professional assistance is greater than ever. The TV archives are filled with an unprecedented variety of formats. In contrast to Germany, North America has a large number of legacies: the stocks of two-inch and one-inch video tapes are in the tens of thousands and have already become largely obsolete because they were not transferred to new storage media in time. On the one hand, this is due to the underdevelopment of an awareness of the specific problems of magnetic storage media and a consequent lack of foresight in the first three to four decades of regular television broadcasting operations. On the other hand, however, this is due to an archive practice that operates breathlessly at the pulse of production rather than with a required overview, which in the past was not always able to decide independently on its primary archive medium and had to orientate itself to the respective formats dominant in broadcasting operations, as archivist Kathy Christensen (CNN) points out:

> Some archives have more control, I guess, over the formats they use. But, basically, whatever the production environment will be, it is going to be our format. So, in that sense it is what it is and then what's true for any archive is that over time that becomes an obsolete format and the ability of play it back is reduced because the machines are old. (Kathy Christensen, CNN)

In addition, there is an uncompromising patronage of television archiving by the industrial specifications with regard to the available storage technologies. The audiovisual history of all conceivable spheres of life and society is as dependent on the deliberate maintenance of stocks and storage material durability as it is on the innovation cycles of the electronics industry, whose economic principles have undermined the archive's efforts to reach a uniform standard format. Until now, it has not been possible for television heritage management to make a reliable estimate of which storage format could become established in the long term. "If you've asked me in 1990, the optical discs seemed to be the answer to everything. And now you can't buy one," says Ivan Harris of the CBC Museum. Since profound knowledge of the market does not allow prospective determination of which storage technology will prevail in the future, wrong decisions regarding format selection are pre-programmed, admits archivist Michael Harms (SWR) among others. Thus, television archives feel like they are in a latent state of emergency, constantly confronting them with the question as to which of their assets need to be left behind in the race for information technology innovations:

> [A]ll the moving image archivists are wondering the same things: What do we do, where do we go, and how will we realize it? There is this market that pushes new technology in the direction what consumers want and you have a hard time to catch up. As an archivist that's what you end up doing: Playing catch up with what you already have. (Ruta Abolins, Walter J. Brown Media Archives and Peabody Awards Collection)

A satisfactory solution to the problem of long-term stability of audiovisual storage media, despite technical progress, is still to be found today, albeit not outside the realms of technological possibilities, but far too far from the strategies of the electronics industry, which measures its success in terms of quarterly results and sales figures, but not in terms of the reliability of its products in an archival sense. To the conservation authorities, the dictates of eternity apply: Not 5, not 10, not 20 years one wants to be able to rely on the stability of the information carriers, but, if possible, forever: "'Long-term' does not mean that the preservation of digital resources will be guaranteed for five or fifty years, but rather the responsible development of strategies that can cope with the constant change caused by the information market" [28]. The storage medium Digi-Beta, which dominated the television industry at the beginning of the twenty-first century, has been in use in many places for almost 25 years. Archives, partly, still have a rich collection of the predecessor formats Betacam and Betacam SP. They are showing signs of ageing, which makes it urgent to migrate large parts of the archived material, that is, transfer it to new formats:

> The industry has gone from kinescope to 1-inch videotape to ¾-inch videotape to Beta SP and all the other formats that all have their specific problems. It's challenging for the archivist to migrate the astronomical high volumes of material before the original legacy tape deteriorates and you lose access to it. (Mark Quigley, UCLA Film and Television Archive)

Two-inch quadruplex, one-inch type A to C, three-quarter-inch U-matic, Betamax: This brief selection of professional tape formats gives only a glimpse of how often parts of the television heritage were already on the brink of extinction because a new, supposedly more powerful, more stable, and more durable short-term standard referred all previous formats to obsolescence; not only the tape formats are disappearing from the market, but also the recording and playback devices. Whereas material storage media such as texts, photos, or even extremely miniaturized microfilm could also be read with "magnifying glass and candle" (cf. [14: 48])—that is, without mechanical support—electronic audiovisual content always requires a technological device in order to be played back. In 1986, the British recording engineer John C. Mallinson was pessimistic that the mechanical dilemma would soon change:

> The machines are the principal source of the archival problem, not the records, tapes, or the discs. And it seems unlikely that future machines will solve the problem. Indeed, it seems more likely that they will exaggerate rather than solve the problem [30: 152].

Mallinson's fears have not only come true but have reached proportions that put the highest watchmen of the television heritage on alert: Archives such as the NAVCC store old equipment up to the ceiling. What looks like a museum repository is in reality an indispensable stock of spare parts because the antiquated machines can only be repaired by slowly dying specialists if they can still be put into operation.

> [W]e've been scouring the earth looking for 2-inch and 1-inch machines and we need people who know how to operate them. My hope is that we can find some retired guys who can come in and work on some of the machines teaching younger people. (Mike Mashon, Library of Congress)

Knowing how to operate these bulky devices is often the only way to explore and evaluate programs from the first 10 to 15 years of television in the 1940s and 1950s and, if necessary, migrate to new tape formats to ensure their survival. When individual archivists talk about their old holdings, one can sense the auratic effect of the early legacies of the programs' operations, which are on the verge of becoming useless because no one is able to play them:

> The 2-inch tapes: big cases and very heavy, the worry with those is that the men who know how to do the transfers and treat the old machines and get a useful copy out of an old tape that is deteriorating: these people are getting older, they are retiring. The long-term concern

is that soon there won't be anybody who can do that anymore. (Margaret Compton, Walter J. Brown Media Archives and Peabody Awards Collection)

Even if the unstable magnetic tapes should survive, the responsible archivists are still far from being able to breathe a sigh of relief, as the shelf life of playback devices after their production has been discontinued due to the lack of spare parts, and often falls far short of that of a car. Archives, therefore, only regard it as a small stage victory in the race against material oblivion if they can at least secure their stocks of one-inch and two-inch tapes from the 1950s and 1960s by gradually transferring the consignments to newer formats. Despite all efforts to find a permanent storage standard, however, migration is expected to remain the only constant in the daily life of a television archivist, as Axel Bundenthal of the ZDF points out:

I think there will never be a common standard. We assume that we will continue to have to transfer regularly to new carriers in regular cycles, which unfortunately are becoming shorter and shorter (was it initially 15 years, then 10 years, today it is only 7 years). (Axel Bundenthal, ZDF)

The costly present-day preservation projects are often planned for several years and can only be partially implemented by public institutions without a desirable financial backing: "We do what we can: A lot of the kind of formal preservation efforts that we have in television are based on the transfer of 2-inch videotape to Digi-Beta and Beta SP. [...] We don't have like a whole programme to transfer everything to one particular format," says Dan Einstein from the UCLA Film and Television Archive. According to the fundamental tenor of the respondents, content migration can only ever be an intermediate step in the fight against time and the acceleration of technological innovation, since even when a preservation project has been completed, the new carrier format threatens to become obsolete.

8.5.2 *The Digital Future: Forced Paradigm Shift or the Solution of All Problems?*

In the search for a way out of the 'hamster wheel' of a constant transfer of archive holdings from one carrier to the next, digital information technology gives a glimmer of hope for a solution in the fight against deterioration and rapidly changing format standards: In the encoding of audio-visual analogue signals into digitalisates, i.e. binary strings without a physical reference value, most preservation authorities see the future of archival culture. The curse of obsolescence seems to be much less threatening in the case of powerful computer technology than in the case of physical preservation instruments threatened by decay. Professional tape formats have been in use since the early 1990s, storing digital signals via magnetic tape recording. Digital Betacam by Sony, or Digi-Beta for short, became the most widely used storage medium in the television industry. However, it is the proliferation of computer-based storage solutions that appears to be an attractive alternative, which overcomes many of the current conservation problems and which prompted television heritage management to ask whether it should completely dispense with archiving video tapes in the future.

Between 2004 and 2007, a new era of dematerialized television storage began in each of the broadcast archives surveyed as broadcasting operations introduced new digital production technologies. However, only a few institutions dared to turn to the digitization of their existing tape-stock at an early stage. An exception is the Vanderbilt Television News Archive, which digitized its entire collection in a concerted effort over a period of only 4 years. The general mood of optimism shows that this will not remain an isolated case: None of the archive managers surveyed ruled out digitization for their own collections; rather, there is a very positive expectation of the archive's possibilities for file-based management of the television heritage. The television archives also had no choice: If they close off the changing imperatives in electronic media production, there is a much more devastating threat of the destruction of archival assets. Not only will the problem of obsolete data carriers and thus the preservation dilemma continue to intensify, but also the archival sovereignty over new carrier formats is at stake.

Concerns and insecurity are mainly based on the handling of genuinely digital documents, the preservation of which has so far not always been a matter of course within the remit of the broadcasting archives. John Koshel of NBC remarks that the term 'archive' is claimed by increasingly different functionaries within the television business and has thus become more indeterminate: In digital broadcasting, everyone has an "archive server," which is why some of the archive management has been transferred to the broadcasters' IT departments. This has certainly profound consequences, if there is to be a further division of archive and archival-like tasks, for the authority of trained archivists in the administration of television broadcasting according to professional standards. In order to remain relevant and to distinguish itself as an indispensable authority for the preservation of televisual works of any aggregate state in opposition to the demands of newly grown productive units when allocating the administration of heritage, television heritage management is faced with the challenge of using information technology whose archival qualities are still under development.

All the more urgent is the question of a reliable long-term preservation, which has culminated since the advent of computer-assisted information processing in the debate about the pros and cons of digitizing the analogously stored cultural heritage. The digitization of archival and collection material is regarded by archivists as an indispensable measure to protect documents from decay, which in some cases is unavoidable, on the one hand, and to improve accessibility or—for example, in the case of particularly sensitive or already attacked documents—to make them possible in the first place. Based on Enzensberger, it is true that what Manfred Osten preferred in his essay on "The Robbed Memory" [36] should not be referred to as apocalyptics and evangelists, who each take opposing, either extreme negative or advocating positions on this issue, but rather in a pragmatic manner from theoreticians and practitioners. The theoretical examination of the stimulating topic of the dematerialization of historical heritage treasures in cultural studies is still dominated by the predominance of writing in Europe, by which the archival term "orientates itself at the vanishing point of alphabetical texts and paper formats" [15: 132]. It is undisputed that the digital encoding of analogous materials leads to a change in the consistency of the memory storage and thus also of the memory spaces postulated by Aleida Assmann (cf. [2: 411]). Electronic memory contents are inviolable: What was previously a conglomeration of different cultural techniques with all senses is now a mere code consisting of two digits—0 and 1. The creation of a material image carrier, on the other hand, requires different materials: magnetic tapes, plastic, silicon, etc.; as a finished product it can be grasped, played back, stowed away. The digitalisat is only a temporary glow of transistors: the "flat, cold, glassy glare of a computer screen" [34: 70].

The fleeting television signal, which has materialized on film or tape, promises to become just as immaterialized again through its decomposition into the digital code as it did during its transmission. As an ethereal medium, television itself lacks any materiality. Nevertheless, a culture of preservation has developed in its favor, which is based on the principles of the preservation of the written word: As such, the television program no longer exists in the archive, but is rigorously broken down into its individual components, which in turn are inserted into the shelf order as cassettes. The archives, with their traditional conventions of preservation, are now faced with a decision as to whether their proven methods and working principles can be used to grasp dematerialized records or whether they have to apply completely different standards to archiving in view of the volume of digitization and the composition of digital storage. For example, an entire administrative apparatus, which is responsible for the permanent safeguarding of the television set, faces a paradigm shift that cannot yet be overlooked in its consequences:

> Apart from examining the skills required of the moving image archivist to deal with the new infrastructure of digital technology, however, we must also examine the economic, political, cultural, and philosophical ramifications of building the digital archive. With our current emphasis on the artifact, the moving image archive field has invested heavily in materiality and in concepts of evidential value: originality and authorial intent, authenticity, fixity of the object (and information contained within it), and the aesthetics of the format itself. Our activities, particularly preservation and restoration, reflect these biases. Much of our presupposed framework will be rethought in the digital era. [20: vii]

Nothing is more dangerous for the cultural fundus—as the development of the television heritage management has shown—when the archive does not know how to follow and yet defy media change at the same time. Like never before, preservation agencies are held responsible for being technically up-to-date, but at the same time they do not want to lose their character as a reliable haven of calm in a relentless storm of media innovation and at the same time stand up for continuity in the preservation of cultural assets, which is becoming increasingly difficult to guarantee in the face of growing uncertainty about the archive qualities of the available storage technologies. In some places, this may explain the reluctance or surrender in front of the quantity of archival assets that are considered for digitization: In the CBS news archive alone, two million video cassettes and almost 17,000 km of film material had to be encoded in order to achieve the digital turnaround.

Despite the continuing fall in prices for storage solutions for private consumers, the digitalization euphoria in the professional archive sector is also being curbed by horrendous costs associated with retroactive digitization of archive holdings, as Axel Bundenthal (ZDF) explains:

> It should be understood that this is not a cheap process, that the storage facilities are complex, require administration and that safety precautions must be taken, because a robot can burn down or be damaged by a water pipe breakage just as much as a classic rack storage system. Or due to serious software errors, data can be difficult to reproduce or stored incorrectly. No, we are in the development phase, and it seems to me that we are still too euphoric. (Axel Bundenthal, ZDF)

A certain "lagging behind" (Hans-Gerhard Stülb, German Broadcasting Archive) that characterized television heritage management for many years is also the result of mistrust of the promising remedies offered by the digital economy to finally find a solution to all analogue capacity and durability problems: None of the archive specialists surveyed can yet clearly state what the actual benefit of digitalization promises for the long-term preservation of program material. For this reason, some institutions such as the Paley Center for Media cautiously proceeded and did not want to immediately break away from their previous archiving principles. What is described by Doug Gibbons as a transition phase, which is primarily due to the large volume of digitization and entails a juxtaposition of analogue and digital systems, is thus also an expression of the balancing act over the conceptual divide between analogue and digital preservation strategies in which television preservation is perceived. The parallel maintenance of analogous archival efforts to simultaneously explore digital possibilities for long-term preservation does not make archival work any easier, but has led to a considerable increase in the workload.

However, it is only possible to postpone selection decisions to a limited extent on which old stocks are to be encoded. Rather, evaluation conflicts are the order of the day. In institutions such as the UCLA Film and Television Archive and the University of Georgia's Media Archive, where it has become accepted that tapes in particular that are threatened with decay must be digitized in order to continue to advocate the most comprehensive possible preservation of the holdings, a decision is made elsewhere with regard to the qualitative benefits of the collection solely on the basis of content:

> First we had the priority that the most endangered programmes should be digitized first. These decisions were made at the very beginning and abandoned very quickly because we realized that endangered programmes are not necessarily quality programmes. Like I said: It's not about to have the oldest collection, but about having the most accessible and educational one. (Daniel Berger, Museum of Broadcasting Communication)

In the broadcast archives, in turn, it is almost exclusively the needs of broadcasting operations that determine which broadcasts are to be digitized, as John Koshel (NBC) and Laurie Friedman (Fox News) describe, among others. The NBC News Archive is also following a so-called hierarchical storage management system for the digitization of its assets: All those tapes that are most urgently needed in broadcasting operations and cause the highest archiving costs on site at the broadcasting center in Rockefeller Plaza in Manhattan were digitized first. "The real driver behind the digitization efforts here—though there are several—are the production needs," says Koshel. According to Friedman, the program archive of Fox News may contain permanent holdings that, for the same reason, may not be digitized at all due to the lack of demand from broadcasting operations: "We don't go backwards, we just gonna use what we need and digitize it as we need it. So, it is possible that some assets won't be digitized at all." The resulting danger of "silent deletion," as Heiko Kröger (NDR) describes the neglect of obsolete stocks, concerns essential areas of programming in view of the extensive archive holdings of the television broadcasters. Nevertheless, the long-term goal of a comprehensive digitization of the collections is propagated, also in order to be able to keep up with the technical development in the long term. The digitization trend is accompanied by ambitious goals and hopes by the surveyed experts, which can be summarized as follows:

- Universal and permanent usability due to uniform data structure
- International standardization and alignment of conservation strategies
- Compression-free archiving in best image and sound quality due to growing storage capacities
- Loss-free and automated migration of audiovisual content
- Optimized security through decentralized storage solutions

If fears of loss appear absurd at first sight due to the exciting new and above all large, that is, capacity-rich storage potential of digital technology, the well-known problems of the analogue era prevail: How can digital mass storage be operated for as long as possible and how can file formats for audiovisual media content still be played tomorrow? In short: How to ensure "to stave off the twin evils of bit rot and technological obsolescence" [27]? The crux continues to lie primarily in the far from trivial problem of preventing the mass uselessness of audiovisual archival material (cf. [4]). Even digital formats do not promise long-term security:

> When you take care of acetate film and take it out a hundred years later, you don't have a problem with it. But when you store something digitally, you might have a corrupt file after two or three months. (Mark Quigley, UCLA Film and Television Archive)

Basically, the more compact and powerful a storage medium is, the greater is its susceptibility to damage. "All archivists are waiting for the big digital fallout where there is some big loss of digital information," Ruta Abolins (Walter J. Brown Media Archives and Peabody Awards Collection) describes the fears in the audiovisual archive sector. While CD-Roms or DVDs were initially classified as a serious archiving option with the prospect of over 150 years of stability, these assumptions have quickly spread in the digital domain as well, which applies to all audiovisual content carriers right up to servers, that is, central computer systems; their service life is extremely limited and is affected by all sorts of disturbing factors from the environment (see also [5: 51]). Digitization does not in any way relieve the memory organizations financially; rather, regular updates of the machinery for digital data management are connected with "huge investment sums", as Hans Hauptstock of the German public regional broadcaster WDR explains, also because of the risk of failure of individual storage devices, which means that complex security systems have to be used.

An even more obvious problem lies in the readability of digital copies, which are stored at a certain time in the most efficient format for archiving. The flood of digital video file formats challenges the development of standards even further. While physical carrier formats can exist on the market for at least up to 10 years or more and thus permit a certain degree of predictability in the migration efforts of large portfolios, file formats are assumed to have an obsolescence frequency of about 2–3 years. In addition, there are no format concepts for the multimedia or even interactive content of Internet television, which has extremely gained in relevance. The unimpaired reproduction of digital applications requires open software technologies that are reliable across platforms as well as from a time perspective, and which enable archival standardized processing despite the fact that isolated solutions are often incompatible with each other. Archival preservation of genuinely digital content from the network is complicated by the lack of a multimedia standard that describes "all media forms of digital objects" [12: 39]. Each media form is still stored separately in different file format standards:

> What is there to define as an object? The website, the individual document or the integrated elements? That is extremely difficult. The law on the German National Library has also been adopted a little bit covertly, in other words secretly, without talking to the actual content producers. We deal with the problem without having a conclusive solution to it. (Axel Bundenthal, ZDF)

There is undoubtedly still too little knowledge of the pitfalls and side effects of digital long-term archiving, which is still in its infancy. The fear of total loss has also spread among television archivists: Steve Bryant (BFI) sums up the often fundamental mistrust of the pitfalls of each new technology in words: "We are responsible for the preservation, so we hesitate because we simply don't know whether we should trust newer forms of archiving." Concerns are still warranted, however, as the complexity of audiovisual data is much bigger than that of text: Digital recordings may be fragile, but there are already system solutions in place that are also used by the television administration to prevent damage. "This digital data can only be preserved if it is kept in constant motion. It's like the juggler with his plates: When he stops juggling, a plate falls off," says Peter Schwirkmann (Deutsche Kinemathek) describing the requirements for digital mass storage, whose greatest advantage is that data can be migrated without human intervention and without loss. Some of the interviewed experts are concerned about the increasing reliance on automated archiving solutions: "[Y]ou better have the original asset and you got to take care of it while enjoying the new technology," says Doug Gibbons (Paley Center for Media). Even if there are no long-term alternatives to digitization, in Gibbons' opinion material backup copies in the form of separately maintained master tapes will remain necessary in order to avoid a complete commitment to the insecure risk of digitization. Axel Bundenthal (ZDF) agrees:

> In my opinion, this problem can only be solved by regularly copying it to the current software and format versions as we operate them. [. . .] At the moment, for example, all the material we have in the digital archive is still available on a classic medium in parallel. In the future, we will also have to convert files into the respective current formats again and again in order to preserve the archived archive. (Axel Bundenthal, ZDF)

This approach is not considered worthwhile everywhere in view of the effort involved: As Mardiros Tavit (ProSiebenSat. 1) and Susanne Betzel (RTL) explain, once digitized broadcasting material is to be preserved only in file form. The playout of files on physical media cannot really be a solution in the long run in view of the growing amount of digitally produced archive material. The simple and lossless transfer and duplicability of digital content rather focuses on a decentralized archiving variant that could offer protection against unforeseeable catastrophes such as the major fire in the facilities of Universal Studios in California in June 2008, which destroyed more than 40,000 video tapes and film reels (cf. [11]). Although the management reassured that it was only copies and that no unique specimens had been lost, such incidents remind us of the destruction of the ancient library of Alexandria. Although such horror reports are rare, it happens again and again that television material is damaged or destroyed by accidents, as is also the case with ZDF, where a lost New Year's Eve rocket in Berlin's capital city studio at the turn of the year 2003/2004, according to archive director Axel Bundenthal, damaged the archive rooms, but did not destroy material. However, four and a half years earlier, in the summer of 1999, arson had caused considerable damage to the former ZDF state studio in Berlin:

> One night a fire broke out there, and some of the stocks were destroyed or severely damaged by the extinguishing efforts. However, ZDF is insured and has been able to reconstruct almost all of its broadcasting material. We even have a charred cassette to remember, which almost looks like an art object. The cassettes were processed, cleaned and transferred. Only a smaller stock was irretrievably lost. [. . .] We are insured against certain damages, but of course—assuming the original total production value—we are not insured in the billions. (Axel Bundenthal, ZDF)

Through the Internet, the institutional archiving of digital television works offers the opportunity to use the establishment of distributed storage facilities not only to achieve effective protection against local disruptions, but also to promote the deepening of cooperation and close networking between partner organizations. The Internet Archive, a California-based nonprofit organization, shows what the future of a new archive paradigm could look like if it is consistently thought through under the sign of digitality: The pioneering initiative for digital web archiving stores the data it collects at its headquarters in San Francisco, as well as at other server sites in Europe and Egypt, to minimize the risk of loss through duplicates: "If there's one lesson we can take from the (destruction of the original) Library of Alexandria, it's don't have just one copy," the founder Brewster Kahle is quoted in the trade press (cf. [10: 42]). Kahle places particular emphasis on cooperation and decentralization in the long-term preservation of digital copies that are not subject to a uniform archive format but are archived in their original form. By relocating data to different server sites and building a strategic storage network, memory institutions can use Internet technologies to create a digital repository that, through its decentralized structure, provides the highest possible level of security against data loss, provided they are archive-certified solutions that prevent unauthorized access and misuse of data and operate according to long-term preservation standards.

Decentralized archiving using digital storage and network technologies is also becoming more and more affordable and attractive for broadcasters. For correspondents' offices, it has long been part of the standard operating procedure to transmit television programs to the broadcasting center via digital broadband lines: Fully automatic data transfer not only simplifies the operation of news production but is also a promising perspective for archive management. The conceptual design of a decentralized archive structure can also be implemented financially for the first time, provided that its advantages are recognized; after all, the smooth transmission of archive content via the Internet can also tempt users to dispense with additional locations in order to keep digital copies available: "So if we leave the digi-band and work with files, the file doesn't care if it is in Bangalore in India or in Munich or Berlin on the server or on a disc. We have the file once, and we can move it back and forth" (Mardiros Tavit, ProSiebenSat. 1).

Digital technology and the Internet alone are, therefore, by no means responsible for a possible storage crisis, but rather for the ensuing consequences for archival practice. The Australian media scientist Belinda Barnet represents another interesting and constructive counter-argumentation to the derivation of a memory crisis through the change in the consistency of storage forms and the development of the Internet: Tying in with Derrida, who assumes man has a primordial desire to preserve his heritage (cf. [13]), she sees the global network as a prime example of a distributive archive, because it was originally intended to be a jointly fed and used resource for scientists and engineers [3: 225]. Barnet sees not only the rapid further development of communication technologies as a decisive driving force for the Internet, but also the underlying understanding of communication as an outsourcing of thought models and a related global "archive drive" [3: 231]. In other words, the Internet was only made possible by the motivation to share and preserve knowledge with one another with the help of a decentralized repository (cf. [26: 262–312]). Barnet also makes the assumption that there are serious deficits of knowledge in contemporary Internet use, in that the recipient digs down more knowledge than registering the actual depth of knowledge [3: 217]. However, the virtualization of memory content through the potentially possible participation of everyone within the global network leads to personal memories becoming nodes within the net of cultural memory [3: 222]. Osten interprets the "copy-and-paste" principle of information circulation on the net as a fertile, memory-relevant habitus: "A long-term survival of memorabilia could therefore be ensured at least potentially by a global ubiquity of digital information clones. This means that the respective information would have to be geographically distributed worldwide through its digital mirroring". [36: 88] The widespread habit of downloading media content from a website and republishing it on other sites such as discussion forums is undoubtedly a copyright infringement. Nevertheless, it cannot be denied that the technological principle of decentralized mirroring of memory contents reduces the risk of loss of the individual digital image. This is also in line with everyday experience that once published on the net, content might quickly be disseminated and then can no longer be removed from the world by simple deletion, because it is not necessarily possible anymore to trace who downloaded it and stored it elsewhere in the network. Memory contents thus become nomads that only remain in motion through constant migration measures, are constantly restless and change their storage location or settle in more and more places at the same time (cf. also [22]). However, this ultimately uncontrollable phenomenon does not release the responsible preservation agencies from the task of finding reliable ways to secure digital security in the long term, also in order to solve legal problems.

8.5.3 *"Spread the Burden": Cooperative Solutions for an Uncertain Future*

The insurmountable dilemma of television preservation of a short life expectancy of technical standards based on the varied history of electronic technologies makes it necessary to guarantee cross-platform continuity in the storage of digital copies and their stability. In view of the high cost of tape-based migration projects, the time is pressing for the development of digital solutions that will have a lasting positive impact on the long-term preservation of television program material. In order to be able to cope with volatile market trends and always be in a position to ensure reliable care and the unproblematic use of managed cultural heritage, there is no way around more cooperative conservation strategies, not least in order to avoid falling behind in the adaptation of digital storage formats. "We want to spread the word, the knowledge and share what we have to the extent we can": What Chuck Howell (Library of American Broadcasting) describes as the overarching goal of his institution is at the heart of a new self-perception of television heritage management, which is being widely disseminated in the course of the digitization trend and does not stop at even archivists acting with commercial impetus. Supposed competitors thus become allies in the fight against the digital sphere's convulsions. Although the exchange of information is mostly (still) limited to personal contacts at association level, where archive representatives inform each other about promising or failed backup strategies, no archive will be able to afford it in the future to renounce the experience of the others.

The fact that the televisual archive community is moving closer together to face new challenges is particularly beneficial for smaller and nonprofit archives, which depend on the bundling of forces and benefit from the general insight that knowledge and technology transfers as well as joint projects benefit all those involved in a digital infrastructure, because even the weakest partner can make a contribution: The more (institutional) actors agree on the use of a technology, the greater the chances of testing it more effectively and reducing the otherwise loss-making trial-and-error effect, the more likely it is that a standard will be formed. However, financing remains the cornerstone of such cooperation: "It would be helpful to share information about practices and technology and storage and where the industry is going, where digital formats are going [. . .], but it always comes down to funding," explains Daniel Berger (Museum of Broadcasting Communications), also with a view to the possible entry of larger institutions into the protection of threatened material in institutions that cannot afford proper archiving—"to spread the burden with a broad cooperation," also on an international level.

In the opinion of several surveyed experts, archive institutions in Germany and North America should, regardless of their commercial or nonprofit orientation, assume at least partial responsibility for threatened foreign television program traditions. As museum advisor Barry Lord explains, in large parts of Asia and on the African continent it has not been possible to establish associations or other organizations in order to better protect the cultural heritage there on a cooperative level. According to Lynne Teather (University of Toronto), the lack of willingness to donate represents a further threat. Especially Africa is a thorn in the side of the interviewees:

> You could work with other organizations in concert on a world-basis to try to preserve the damaged heritage of countries that are in constant danger. Think about the archival work in Africa! In terms of audio-visual archives there is literally nothing in Africa at all! There is a little bit in Simbabwe at some time, but that was a British guy and he might been kicked out by now. There is a whole continent that is like a desert in terms of that audio-visual archives are concerned. (Sam Kula, Consultant)

Researcher Henry Jenkins (MIT) believes that it is above all the major television museums that have a duty not only to take more care of the collection of foreign television programs, but also to make an effort to preserve them locally should this part of the global television heritage be threatened. Gerlinde Waz of the Deutsche Kinemathek and Dietmar Preißler of the Haus der Geschichte der Bundesrepublik Deutschland consider this to be respectable, but for cost reasons hardly feasible and agree with Daniel Berger, who sees in an international alliance of preservation institutions the only way to construct a sustainable basis for such a request. Nevertheless, Peter Paul Kubitz (Deutsche Kinemathek) emphasizes the need for an initiative to this end and sees a commitment by supranational organizations such as UNESCO as a good opportunity for implementation:

> I believe that this would have to be linked to UNESCO in some way. [. . .] This isn't a completely new land. The UN has awarded the World Heritage label in all possible areas: Here it is a glacier, there is a desert or a historical old town. This is basically the same for certain audio-visual sectors. It is not only about 'Memory of the World', but also about preventing the disappearance of programmes that one doesn't know what they will mean in twenty, thirty or forty years—where the loss is probably not even noticeable because one doesn't know about it at all. (Peter Paul Kubitz, Deutsche Kinemathek)

On the other hand, in individual cases, development aid has already been provided by broadcasters for a long time, which is coordinated by the international television archive associations International Federation of Television Archives and the Association of Moving Image Archivists. It is primarily representatives of public or public broadcasting archives who believe that they have a responsibility to look beyond the national horizon and to provide development aid. Michael Harms (SWR) speaks of a colleague who, with the support of the broadcaster, undertakes business trips to give lectures to Malaysian archive staff and also offers targeted training in other developing and emerging countries in order to help them obtain the program material of the respective television broadcasters. Geoffrey Hopkinson (CBC) also recognizes the task of promoting a more educated archive awareness and knowledge in certain parts of the world, where deficiencies in television retention are identified:

> I was asked to go to the Bahamas to look at their archives and make recommendations for preservation and moving forward. My counterpart in Montreal just came back earlier this year from Algiers to look at their archives. This is part of our former vice president of English television who retired and installed a need for public broadcasters to help third world countries in terms of their television preservation, production, whatever. We've started to go on with that. [...] What I really would like to do is, once I have finished my preservation, to look at others and perhaps work with them. There are very few 2-inch machines left in the world, so maybe we can preserve somebody else's collection of 2-inch, then we would start looking at that—for example on the Bahamas. (Geoffrey Hopkinson, CBC)

One of the central findings from the transformation processes within archival work, which changes from analogue to digital preservation strategies, is that the perspective goal of a general strengthening of the televisual archive sector in the concert of the diverse levels of actors in the media industry promises long-term—also monetary—benefits. "Preservation of digital content must be a collaborative effort that involves the professional archivist, the technology expert, the user, and the creating and producing entity" [23: 98]. Joint efforts are paying off, as evidenced by initiatives such as the US Preserving Digital Public Television collaboration program between the public broadcasters WGBH, Thirteen/WNET, Network PBS, and New York University: The initiative, which was launched in autumn 2003, acquired millions of US-dollars in funding through a joint effort and carried out fundamental work on new technical problems, selection, and evaluation questions as well as development methods with the aim of developing scenarios for the long-term archiving of digital television productions. The best-practice models developed are a substantial contribution to the development of presentation strategies in the entire field of television heritage management. The project also gave rise to the "American Archive" initiative, which aims to make the archive holdings of public broadcasters in the United States digitally accessible to a broad public. The national funding program places high expectations on the willingness of small local television stations to cooperate, which are attested by a partly underdeveloped archive awareness, but which would only contribute to the success of the project through their broad distribution:

> WGBH and Thirteen create 60 percent of the national programming seen on public television, but the American Archive has to include small stations as well as independent and minority producers in both television and radio. [...] I consider it really important to represent those community producers. And they're not going to be brought to the table unless they have advocates. It's not just Great Performances and Nova that need

> to be preserved; it includes all these other parts of public broadcasting—local stations, independent and small producers, who represent the heart of our rich collective programme sources. (quoted after [1])

In addition to such cooperation efforts, which, after a long history of individualism, will allow the preservation of television to develop a lobby status that will now be able to promote its claims and needs more effectively toward politics and the electronics industry, there is also the rare opportunity to build bridges between the previously strictly separated cultural heritage management agencies and, above all, the institutions of document archiving with those audiovisual traditions.

Archives, libraries, museums, research disciplines, and all mass media sectors of the economy, including the literature industry, are challenged by the digital era and have to cope with its imperatives when it comes to safeguarding the cultural heritage. Nevertheless, representatives of television archives complain that the rest of the community is unwilling to engage in dialogue with institutions committed to preserving cultural traditions. In Germany, for example, the institutional network "Nestor" ("Network of Expertise in long-term STOrage and availability of digital Resources in Germany") is highly engaged in the development of long-term archiving methods for digital resources. The project partners include the German National Library, the Federal Archives as well as several university libraries and the Berlin Institute for Museum Research. Due to the focused expertise of the participating institutes, moving image archiving is inevitably pushed into the background or is almost completely ignored, which is regarded as an essential shortcoming of the initiative. It still seems utopian that there will be greater interdependence between the cultural preservation organizations of different forms of media. Cooperating with document and AV experts is essential, especially with regard to the recording of multimedia network publications (see also [21: 302]), as digital content in any media format has the same advantages and problems. However, this does not only require common goals: Mutual respect for the cultural value of the respective traditional areas is also a basic prerequisite for successful collaborations (cf. also [6: 223]).

Television archives have valuable experience that can also benefit the rest of the archive landscape. Small association sections such as Division 7 of the Association of German Archivists' Media and Documentary Archives could evolve into "think tanks," which, with their practical experience, can also be used as a guideline for the problems of yesterday's, today's, and tomorrow's digital preservation of cultural heritage and communicate possible solutions. However, as long as there is no mechanism for linking the different archival areas under the auspices of digitality promising cooperation approaches will depend on the insight of individual actors that progressive media change will also force archives to rethink their genre-specific affiliations and open up to the convergence movements in order to preserve their holdings sustainably and at the same time be able to record future technological variations.

8.6 Conclusions

The television heritage appears to have an open wound in a deeply mediatized society that only recently discovered and learned to appreciate and appropriate the richness of its media history. The fragmented archival stocks, representing the versatile programming history of television, can only stimulate a blurred memory of the original broadcasting contexts, because—in most cases—they can hardly be reconstructed. There is no doubt that television has not made it easy to raise awareness of its historicity and the need for conservation and care measures: In its seductive presentism, the singular linearity of its broadcasting process, it seemed barely tangible, not least because there was and still is a lack of resources to take care of its preservation in addition to the loads of ever new programming material. The unbridled televisual flow of programming flooded not only its recipients, but also those who were responsible for it. It was only late that measures were taken to put a stop to the memoryless actions of the television industry. In the meantime, no more cassation patrols have been rushing through the archives for a long time, erasing important testimonies of television history and permanently handing them over to oblivion. Nonetheless, as has been shown, the situation has only partially eased and has only apparently given way to an eager zeal for conservation driven by the commercial incentive. Even today, television heritage management still has difficulties in getting hold of the medium entrusted to it and in guaranteeing the long-term security of its works. In summary, the following problem areas in the preservation of television programming traditions can be identified.

The endarchival expertise perceived by the broadcasters in North America and Germany from the very beginning of television operations had not only significant losses of archival assets but also repercussions for the basic prerequisites for the maintenance of television programming heritage; the independent management of conservation efforts has led to a highly uneven adoption of storage technologies. The divergent format decisions, which were taken separately by each broadcaster and production company due to a lack of coordination within the television industry, also meant that television heritage management was not able to develop into an influential lobby that was and is able to assert its interests vis-à-vis the electronics industry with a unanimous vote and a conformal security concept. Even today, archives and collection centers are still struggling to shorten innovation cycles in format development and have difficulties in keeping up with the progress made in other sectors of the economy. The necessary arranging with the market conditions in the development of storage capacity is a central obstacle to the efforts of preservation agencies to achieve an eternization and increases the dilemma surrounding the management of television heritage, which means that it is not possible to free itself from its mechanical dependency on account of its essential object of tradition.

The distribution of roles between the television industry and public institutions in the preservation of television traditions has not changed in principle: Broadcasters and production companies maintain their endarchival competence and have successfully resisted attempts to influence the management of their archive assets. It is particularly risky that, due to business considerations, inventory adjustments are made under the sign of cost-efficiency, where duplicates are regularly discarded. Public institutions must rely, well or badly, on the reliability of archiving measures by the television industry, since in North America and Germany no control or sanction mechanisms have been implemented to ensure, for example, that in the case of a cassation of archival stocks the assets are offered to an independent collection facility before it is destroyed in order to decide on its historical value for the general public. As long as in the absence of infrastructure arrangements it is not possible to clarify where and to what extent archival stocks have been threatened or already been damaged and to what extent, the absence of such a legal deposit and the self-administration of singular conservation institutions must be assessed critically.

The volatility of the electronic television signal and the resulting fundamental problem of its difficult handling in the conservation context not only require increased resources, but also an appropriate sensitivity to the technical pitfalls involved in maintaining and dealing with old recordings. In the past as well as today, albeit under different circumstances, there is reason to doubt that both conditions will be sufficiently complied with. Neither could the fundamental problem of the scarcity of resources in archives be satisfactorily solved, nor can the deadline pressure of broadcasting operations permit only in exceptional cases a cautious use of the traditional material, which is often still regarded primarily as an object of production and not as a commodity worthy of protection. Thus, on the one hand, the existing gap between ahistorical utilization by personnel from outside the archives and, on the other hand, the arbitrary actions of individual television actors who do not want to hand over their works considered valuable (be it broadcasts or props) or the written documentation of their actions to the imponderables of a production archive could not be completely closed.

While television broadcasters can be attested to a rather defensive approach in the areas of reconnaissance and the determination of loss of tradition, public memory organizations have so far fulfilled this task. The archive work of the broadcasters, which is characterized by pragmatism and image maintenance, obviously prohibits the public addressing of the problem and the admission of omissions. However helpless or limited non-profit institutions may be due to their scarcity of funds for the long-term preservation of television programming material, the example of the "Lost Programs" campaign of the Paley Center for Media and the predecessor initiative of the British Film Institute show how the involvement of public institutions can generate broad attention and make a significant contribution to the development of traditions that are already believed to have been lost. The involvement of all public spheres in the search for evidence of television history, which is always also a testimony to the history of society, cannot be overestimated in its significance for the cultural recognition of the medium as a memory generator and its future preservation. In Germany in particular, there is a lot of catching up to do in this area, which is due in particular to the extremely late institutionalization of a televisual memory organization by international standards.

The paradigm shift in the management of television heritage, which is announced by the increasing digitization, presents great opportunities but also risks for the future of television broadcasting, not because of the changing state of the archive material itself, but because of the unpredictable wealth of consequences in terms of costs, technical requirements, and durability. As much as television archives and museums look for answers to the high obsolescence frequency of tape formats in digital code, it is not possible to reliably determine the long-term risks associated with the conversion of storage to file formats. However, it is already foreseeable today that the hoped-for easements in the automatic migration of program traditions will take years to come. It remains to be seen whether at the end of a long transition phase, which has considerably increased the workloads in many institutions due to a parallel management of the inventories in analogue and digital content management systems, a more advantageous backup solution that meets the requirements of audio-vision better, remains to be seen. So far, the promise of salvation of digitization has not yet been fulfilled from a preservation point of view, at least not yet, since the reconstruction of the required archive structure has turned out to be much more complicated and materially expensive than expected in many places. Small, financially weak institutions in particular can easily fall behind, but even large broadcasting archives are not immune to erroneous decisions in the digitization strategies of their material, so that further losses are threatened because old mistakes could be repeated under the influence of economic constraints.

Similarly, the ethical problems encountered with digitization in securing the integrity of the archived content are also unmanageable. Even if an irrevocable deletion of digital copies is not easily possible, its content can still be altered imperceptibly by the underlying alphanumeric code. Not only the conversion of file structures during the repeated migration of digital data could automatically lead to a falsification of the original content, but it also increases the risk of opening doors and gates for interventions. A digitally stored document can be changed at will, unless it has been saved in a write-protected format, without being able to trace the contents of the document: no etchings, deletions, cut marks or other indications of editing can be found in cleverly manipulated electronic transmissions. Such a risk of manipulation, however, was always associated with every type of media—in varying degrees and with the appropriate skill of the counterfeiter. The fact that forgeries could manipulate historical TV content in the archive's shelter may be an unlikely scenario, but the hypothetical danger of being unable to distinguish between original and forgery in the event of manipulation of digitized unique copies raises the question of whether the television heritage management tends to be guilty of facilitating abuses through the possible early, comprehensive digitization of all audiovisual, written, and photographic television content. In this way, in the event of damage, the already suggestive television picture, which functions so effectively as a reminder cause and replacement, could become a dangerous instrument of historical falsification and, in its audiovisual appearance, could lead to technologically induced memory mutations.

The proverbial pull in one direction has proven to be a useful strategy for the development and management of solutions for digital preservation systems within the television heritage management. Cooperative models, however, are still in their infancy and are largely limited to personal contacts among the individual archivists and collection managers, who exchange information at association meetings but exclude institutional cooperation at archive level due to the entrepreneurial competitive relationship between the television broadcasters. What is more serious, however, is that, in view of the fundamental upheavals brought about by digitization, which affects all areas of documentary work regardless of the sector, communication between the various areas of documentation appears to be so severely disrupted that approaches to solving problems relating to the long-term preservation of cultural heritage in digital form are not discussed together. Thus, the guardians of the paper and those of the audio-vision discuss their respective problems to a large extent among themselves, but without developing necessary cross-media models for the increasingly merging media genres. Who the impulse for rapprochement must emanate from is subordinate: Measured against the high value of the television heritage for the maintenance of cultural identity, the diagnosed reluctance of the cultural administration to cooperate independently of the media to preserve and strategically develop new long-term security systems is an incomprehensible and fatal omission.

In this way, the opportunities and risks of digital long-term archiving are balanced and require a balanced approach to the preservation of the television heritage. Physical access to archival material will be replaced by virtual access to the virtual, facilitated management and availability, often by making irreplaceable digital surrogates, which are often particularly vulnerable to decay, available. With the advancing development of presentation solutions for the management and processing of digitized documents, the advantages of the expanded accessibility of the archive and collection items independent of time and place, simplified and accelerated navigation and retrieval possibilities, as well as the cumulation of different document types in a virtual database environment are the advantages. The digitization projects of the television heritage management therefore always have two things in mind: rescue of threatened film or video tape material and improved manageability and ergo usability for use. The latter objective goes hand-in-hand with the increased demand for archived programming material for broadcasting operations, as well as from the general public, whose demand continues to drive marketing measures in the television industry. Nevertheless, a large part of the television heritage is still almost inaccessible.

References

1. Anon: Digital preservation pioneer: Nan Rubin. In: Library of Congress Digital Preservation Newsletter, November 2008. http://www.digitalpreservation.gov/news/newsletter/200811.pdf (2008)
2. Assmann, A.: Erinnerungsräume. Formen und Wandlungen des kulturellen Gedächtnisses. C.H. Beck, München (1999)

3. Barnet, B.: Pack-rat or Amnesiac? Memory, the archive and the birth of the Internet. Continuum: J. Media Cult. Stud. **15**(2), 217–231 (2001)
4. Besser, H.: Digital longevity. In: Sitts, M. (ed.) Handbook for Digital Projects. A Management Tool for Preservation and Access, pp. 155–166. Northeast Document Conservation Center, Andover (1999)
5. Besser, H.: Digital preservation of moving image material? Moving Image. **1**(2), 39–55 (2001)
6. Besser, H.: Collaboration for electronic preservation. Library Trends. **56**(1), 216–229 (2007)
7. Billington, J.H.: Preface. In: Murphy, W. (ed.) Television and Video Preservation 1997. A Report on the Current State of American Television and Video Preservation, pp. xiii–xxiv. Library of Congress, Washington, DC (1997)
8. Boyle, D.: Video Preservation. Securing the Future of the Past. Media Alliance, New York (1993)
9. Cameron, D.F.: Environmental control: a theoretical solution. Museum News. **46**(5), 17–21 (1968)
10. Chen, A.: Making web memories. eWeek. **23**(44), 41–42 (2006)
11. Cieply, M.: Fire destroys parts of a popular movie lot in California. N.Y. Times, June 2, 13 (2008)
12. Coy, W.: Perspektiven der Langzeitarchivierung multimedialer Objekte. Nestor, Berlin. http://nbn-resolving.de/urn:nbn:de:0008-20051214015 (2006)
13. Derrida, J.: Archive Fever. A Freudian Impression. University of Chicago Press, Chicago (1996)
14. Encke, J.: Tief im Südwesten. Das sicherste Endlager der deutschen Kultur liegt in einem Bunker bei Freiburg. In: Rühle, A., Zekri, S. (eds.) Deutschland extrem – Reisen in eine unbekannte Republik, pp. 46–50. C.H. Beck, München (2004)
15. Ernst, W.: Das Rumoren der Archive. Ordnung aus Unordnung. Merve, Berlin (2002)
16. Farhi, P.: XLI Years Ago, the Super Bowl Was Just An X-Small. Washington Post, February 3, M07 (2008)
17. Feldmer, S.: Nie mehr vom Band. Süddeutsche Zeitung **51**, March 03, 15 (2009)
18. Fiddy, D.: Missing Believed Wiped. Searching for the Lost Treasures of British Television. British Film Institute, London (2001)
19. Fisher, L.: Memories linger but the tapes fade. N.Y. Times, November 28, 9 (1993)
20. Gracy, K.: Editor's introduction. Moving Image **8**(1), vi–viii (2008)
21. Hans-Bredow-Institut: Zur Entwicklung der Medien in Deutschland zwischen 1998 und 2007. Wissenschaftliches Gutachten zum Kommunikations- und Medienbericht der Bundesregierung. Endbericht. Hans-Bredow-Institut für Medienforschung, Hamburg (2008)
22. Hoffmann, N.: Von mobilen Logbüchern und vermeintlichen Ja-Sagern. Das Internet als Ort mobiler Wissenskonstruktion und -subversion. In: Gebhardt, W., Hitzler, R. (eds.) Nomaden, Flaneure, Vagabunden. Wissensformen und Denkstile der Gegenwart, pp. 159–170. Springer, Wiesbaden (2006)
23. Ide, M., MacCarn, D., Shepard, T., Weisse, L.: Understanding the preservation challenge of digital television. In: Council on Library and Information Resources, Library of Congress (ed.) Building a National Strategy for Preservation: Issues in Digital Media Archiving, pp. 84–99. Library of Congress, Washington, DC (2002)
24. Kramp, L.: Weniger ist weniger. die tageszeitung 7982, May 29, 18 (2006)
25. Kramp, L.: The changing role of television in the museum. Spectator – Univ. South. Calif. J. Film Telev. Crit. **27**(1), 48–57 (2007)
26. Kramp, L.: Gedächtnismaschine Fernsehen. Band 2: Probleme und Potenziale der Fernseherbe-Verwaltung in Deutschland und Nordamerika. Akademie Verlag, Berlin (2011)
27. Lavoie, B.F.: The fifth blackbird. Some thoughts on economically sustainable digital preservation. D-Lib Mag. **14**(3/4). http://www.dlib.org/dlib/march08/lavoie/03lavoie.html (2008)
28. Liegmann, H.: Einführung. In: Neuroth, H., Liegmann, H., Oßwald, A., Scheffel, R., Jehn, M. (eds.) nestor Handbuch: Eine kleine Enzyklopädie der digitalen Langzeitarchivierung, p. 7. http://nestor.sub.uni-goettingen.de/handbuch/nestor-Handbuch_01.pdf (2007)
29. Littleton, C.: MT&R finds '54 'Angry Men'. Hollywood Reporter, April 16 (2003)

30. Mallinson, J.C.: Preserving machine-readable archival records for the millenia. Archivaria. **22**, 147–152 (1986)
31. Martin, J.: The dawn of tape: transmission devise as preservation medium. Moving Image. **5**(1), 45–66 (2005)
32. Menke, B.: Bombensicheres Gedächtnis. Spiegel Online, March 13. http://www.spiegel.de/unispiegel/studium/0,1518,611647,00.html (2009)
33. O'Shaughnessy, L.: Keeping Your Collection off Ebay. A little planning can make a big difference for your heirs and your treasures. Bus. Week, June 04, 104 (2007)
34. O'Sullivan, C.: Diaries, on-line diaries, and the future loss to archives; or, blogs and the blogging bloggers who blog them. Am. Arch. **68**(1), 53–73 (2005)
35. Öhner, V.: Konstitutive Unvollständigkeit. Zur Archivierung und Rekonstruktion von Fernsehprogrammen. Montage/AV. **14**(1), 80–92 (2005)
36. Osten, M.: Das geraubte Gedächtnis. Digitale Systeme und die Zerstörung der Erinnerungskultur. Eine kleine Geschichte des Vergessens. Insel, Frankfurt am Main (2004)
37. Pröse, T.: Der Schatz im Schwarzwald. Focus. **11**(33), 104–109 (2003)
38. Roper, J.: The American Presidents. Heroic Leadership from Kennedy to Clinton. Edinburgh University Press, Edinburgh (2000)
39. Shales, T.: Love that, lost' Lucy. Series pilot seen as ratings booster for CBS. Washington Post, April 30, B01 (1990)
40. Snider, J.H.: Local TV News archives as a public good. Press/Politics. **5**(2), 111–117 (2000)
41. Swartz, M.: In search of lost episodes. Pittsburgh Post-Gazette, November 07, F-10 (1999)

Printed in the United States
By Bookmasters